Your Future on the Faculty

Your Future on the Faculty

How to Survive and Thrive in Academia

Joshua Schimel

OXFORD
UNIVERSITY PRESS

Oxford University Press is a department of the University of Oxford. It furthers
the University's objective of excellence in research, scholarship, and education
by publishing worldwide. Oxford is a registered trade mark of Oxford University
Press in the UK and certain other countries.

Published in the United States of America by Oxford University Press
198 Madison Avenue, New York, NY 10016, United States of America.

© Oxford University Press 2023

All rights reserved. No part of this publication may be reproduced, stored in
a retrieval system, or transmitted, in any form or by any means, without the
prior permission in writing of Oxford University Press, or as expressly permitted
by law, by license, or under terms agreed with the appropriate reproduction
rights organization. Inquiries concerning reproduction outside the scope of the
above should be sent to the Rights Department, Oxford University Press, at the
address above.

You must not circulate this work in any other form
and you must impose this same condition on any acquirer.

Library of Congress Cataloging-in-Publication Data
Names: Schimel, Joshua, author.
Title: Your future on the faculty : how to survive and thrive in
academia / Joshua Schimel.
Description: New York, NY : Oxford University Press, 2023. |
Includes bibliographical references and index.
Identifiers: LCCN 2022018771 (print) | LCCN 2022018772 (ebook) |
ISBN 9780197608838 (paperback) | ISBN 9780197608821 (hardback) |
ISBN 9780197608852 (epub)
Subjects: LCSH: College teachers—Vocational guidance. | Teacher effectiveness. |
Universities and colleges—Faculty—Professional relationships. | Mentoring in education.
Classification: LCC LB1778 .S28 2023 (print) |
LCC LB1778 (ebook) | DDC 378.1/2—dc23
LC record available at https://lccn.loc.gov/2022018771
LC ebook record available at https://lccn.loc.gov/2022018772

DOI: 10.1093/oso/9780197608821.001.0001

Paperback printed by Integrated Books International, United States of America
Hardback printed by Bridgeport National Bindery, Inc., United States of America

To Mary Firestone. You were amazing as my Ph.D. advisor, and you have been a wonderful mentor and friend throughout my career. I wouldn't be here without you, and no one has taught me more about how to be a professor and an academic.

Contents

Preface ... ix
Acknowledgments ... xi
Introduction: The Nature of Academe ... xv

SECTION 1. YOUR INDIVIDUAL PATH AS AN ACADEMIC

1. Postdoc: A Postdoc's Job Is to Get a Job ... 3
2. Assistant Professor: Making It to Tenure ... 14
3. Success: Tenure ... 29
4. Thriving in Academe When You Are Not a Heterosexual White Man ... 37
5. Non-Tenure-Track Teaching Faculty ... 48

SECTION 2. UNIVERSITY SYSTEMS

6. University Administrative Systems ... 61
7. Working with the Staff ... 78

SECTION 3. THE NEXT GENERATION

8. Mentoring 1: Vision and Philosophy ... 93
9. Mentoring 2: Specific challenges ... 105
10. Teaching: Being Good while Surviving ... 117

SECTION 4. PROFESSIONAL COMMUNITIES

11. Publishing Ecosystems: The Editorial and Review Process ... 135
12. Who Put the *Peer* in Peer Review: Being Part of the System ... 147
13. Professional Communities ... 160

viii   Contents

14. Resolution: Thriving in Academe   **166**

Appendix 1: Useful Resources   *169*
Appendix 2: Mottoes for Memorable Mentoring   *171*
Sources Cited   *173*
Index   *177*

Preface

When I wrote *Writing Science*, I felt it was my autobiography. This book is even more so—many of the examples are from my own experiences and observations. I offer them as illustrations of themes, issues, and ideas to provide perspectives on how to maximize your personal happiness and professional success as an academic.

As with *Writing Science*, this book is a way to say "thank you" and to pay back (and forward) to the academic world for having given me the wonderful opportunities and career I've had. This book reflects my ecologist's perspective on systems: What makes them work? How do the parts integrate to create a healthy and functional system? It reflects my love of academia, with all its faults, as well as my hope that from my years in the system, I can offer useful advice for new academics, and insights that we can use to make our systems healthier and happier for all.

As with *Writing Science*, I include examples of both good practice and bad. And I use the same policy: Examples I hold up as good practice, I use intact and cite properly—people who stand as role models for being both a good person and a good academic deserve recognition. When I give examples of what I view as poor behavior by others, I avoid names and limit details to avoid leaving "fingerprints" that might identify the participants. My own mistakes, I have no such hesitance about—I've made enough over the years to provide much fodder. We learn from mistakes, but they don't all need to be our own mistakes. I hope mine offers some insight to help you avoid repeating them. Learn from my scorched fingers instead of your own.

I struggled with the book's title. My first working title had been *Surviving Science*. But I quickly realized that was terrible—the book isn't just about surviving, but thriving, and it isn't just about science. The final title was motivated by riffing off my father, who was a psychoanalyst. He'd written two books exploring the deeper aspects of married life, which most of us only figure out (the hard way) once we are married: *Your Future as a Husband* and *Your Future as a Wife*. I'd never read either book, but they were always in Dad's office, and with this book I am trying to do something analogous. The title honors his memory.

My focus is centered from what I know best—I am a tenured natural scientist who works in a research university and is involved in both graduate and undergraduate education—but I try not to limit myself to that focus. I have a joint appointment in the Environmental Studies Program, which spans from geology and ecology to history and ethics; I chaired the program for nine years. I've also served on, and chaired, academic senate committees at both campus and systemwide levels; currently, I'm Associate Dean of Sciences. Between all those activities, and collaboration with colleagues from across campus, I think I have come to understand university life in a broad perspective, which I try to bring to these pages. I've tried to limit my natural bias toward the natural sciences, but I won't apologize for any that shows up.

I hope that what I say will be useful to people who aren't "me" and are in different disciplines, different types of institutions, and different positions. There are patterns and challenges that are universal to academic life. Although our various fields operate differently, the commonalities of being an academic are greater than the differences across fields. In places where areas of scholarship work differently, I have noted that and tried to provide balanced perspectives.

I've also written this book largely from the perspective of American academe. For example, I use American academic titles (assistant, associate, full professor) and I regularly assume the U.S. tenure system. There are, however, more commonalities among nations' academic systems than there are differences. Universities carry out the same functions worldwide: we teach undergraduates, do scholarship, and train graduate students. We have to balance our activities across teaching, research, and service to our institutions and communities. We all deal with our wider professional communities. I've spent time in universities around the world and they all felt very familiar—regardless of whether an academic's title was Lecturer, Professor, or Professeur, and whether you go for tenure or habilitation. Most of the messages, and certainly those in Section 4 about our wider professional communities, will likely resonate even if you are not in an American college or university.

I hope you find this useful and that some of what I offer may help you sort your way through a career as an academic. Good luck.

Acknowledgments

As with all projects of this scope, there are many people who contributed over the years it has been developing—to say nothing of the people who, over my years working in universities, helped me learn how academe works. None of us build our careers single-handedly, and I am grateful to all of you—more people than I can list or remember.

I'll start by thanking the people who have given me functional departments to work in: the business officers who held my hand and kept me on the straight-and-narrow, and the support teams they have led. Jean James was the business officer at the Institute of Arctic Biology at the University of Alaska Fairbanks (UAF) when I was Assistant Professor. Jo Little in the University of California Santa Barbara (UCSB) Environmental Studies Program was my closest teammate for the nine years I was chair—I couldn't have survived without her guidance and support. Finally, Cathi Arnold has held together the Department of Ecology, Evolution, and Marine Biology. Department chairs may come and go, but our senior staff stick around; they lead the teams who give us a functioning university to play in. I am grateful to you all. Equally, I am grateful to the teams these people led, people who have supported me.

The other key staff members who taught me how universities work were my teammates on the UC Academic Senate committees I've served on: Bryant Wieneke (Graduate Council), Kyle Richards (Planning & Budget), and Michael LaBriola (University Committee on Planning & Budget).

Pierre Wilzius, Dean of Sciences at UCSB, is my "boss" in my role as associate dean. Whether I should thank him for inviting me to serve might still be an open question, but no one forced me to say "yes!" He is great to work with, giving me deeper perspectives into higher-level administration, and his staff team is wonderful—people who are all smart, insightful, and lovely to work with. My responsibilities center on academic personnel issues and Shawnee Oren and Kathy Jenquin most closely guide me through my work. They, Joan-Emma Shea (the other Associate Dean), and the rest of Pierre's team (Alex Radde, Dorothy Satomi, Nancy Emerson, Heather Liu, Dave Davis, Ed Blaschke) have been friends and teammates for the last several years. Thank you.

I can never thank Dr. Mary Firestone, my Ph.D. advisor, enough for the influence she has been in my life. Mary has been a role model and mentor my entire career. She showed me what it meant to be an academic when I was a Ph.D. student at UC Berkeley. When I moved back to the University of California at Santa Barbara and became involved in Academic Senate affairs, she was once again a great mentor. I have dedicated this book to her to honor her contributions, not just to my career, but also to her many other trainees; we have become quite the large clan.

There are several chapters that I could never have written on my own—I couldn't offer personalized advice to people who weren't "me." Writing Chapter 4—thriving in academe when you are not a heterosexual white man—terrified me. It was critical to include, but I am all those things! Drs. Asmeret Berhe and Erika Marin-Spiotta each offered valuable and critical inputs, while a number of other people offered thoughts that I wove into the chapter—notably Drs. Debra Perrone, Maria Herrera Sobek, and Elisabeth Holland. I also couldn't have crafted that chapter without the wisdom of Dr. Kerry Ann Rockquemore and Tracey Laszloffy, authors of *The Black Academic's Guide to Winning Tenure—Without Losing Your Soul*. Dr. Helene Gardner gave me much of the material from which I crafted Chapter 5 on non-tenure-track positions. She also read the draft and helped me revise it into its final form. Dr. Linda Adler-Kassner, faculty director of the Center for Innovative Teaching, Research, and Learning offered valuable insights for Chapter 10 on teaching.

Other people gave me specific ideas and inputs that enriched the book. Stephen Heard gave me the line "find someone to ask" in Chapter 7. Lieutenant Colonel Travis Buehner of the UC Santa Barbara ROTC program offered valuable insights into how the military approaches training and career development. Eric Zimmerman has been the student affairs coordinator for Environmental Studies since I joined UCSB and over that time I have benefitted from his wisdom. My brother, David Schimel, gave me the first critical step into the world of ecosystem science, and also the key advice that is an important theme in this book: *Remember who your real peers are*. I also thank Jerry Melillo who took a gamble on an imperfect chemistry major and hired me as a lab tech—that was the key transition that launched me into the direction I've followed.

I have learned an enormous amount about being an academic, how our systems work, and how to function within them from my many faculty colleagues. A few who have contributed most notably are Carla D'Antonio, Oliver Chadwick, and Simone Pulver in Environmental Studies; Mark Brzezinski in Ecology, Evolution and Marine Biology (EEMB); Patricia Holden in the Bren School; and Janet Walker in Film and Media.

I thank my wife Gwen, who supported my writing and academic career—I wouldn't be where I am without her, in both the figurative and literal senses (I moved to UCSB to be with her). My sister, Liz, catalyzed the choice for this book's title. I also thank Jeremy Lewis and Oxford University Press for their support of both my first book, *Writing Science,* and now of this book.

Beyond these, there are more people than I can name or even remember who have helped build my career and teach me how to function as an academic—many, many dedicated staff members at UAF and UCSB, program officers at the National Science Foundation, and the UC academic senate. I am grateful to my students and postdocs over the years who've been my friends and co-workers; you are the people who have done the work and written the papers that built my career. You've taken care of me—thank you.

Introduction

The Nature of Academe

It's interesting I never had any doubts about the music; but the reality of the music industry is something I had to learn.
Maggie Rogers; National Public Radio interview January 2015

When you start playing tennis, you don't imagine there's a whole bureaucracy behind the tournaments and all of that. You just think about winning the cups.
Venus Williams

One of my early Ph.D. students flowed through graduate school easily and painlessly; the vast hours and intellectual challenges associated with being a student didn't faze him. It was a little infuriating—this is supposed to be hard, dammit! But then he got a job as an Assistant Professor, and I finally saw some stress crinkling his eyes. The science was manageable, but combined with writing proposals, teaching classes, mentoring students, and committee work... that, he noticed. This job is hard, but it isn't the core scholarship—the "music"—that makes it so. It's the "music business" that makes it hard—the complex mix of different activities that we have to juggle and that we have to figure out *how* to juggle.

I'd already started thinking about this book when I heard the Maggie Rogers NPR interview on National Public Radio while I was driving to work; her comment summed up much of what I had been thinking. I got trained in graduate school to do *the science*, but not to deal with the science *industry*—we don't get trained to be an *academic*.

This was something that had periodically perplexed me. When I was in my first year of graduate school, Mary Firestone, my Ph.D. advisor, said she wanted to spend half her time in the lab. She was still an Assistant Professor, yet already didn't do her "own" hands-on research in the lab (though she helped with field work—tThank you Mary). That seemed odd to me. Mary's Ph.D. involved some of the most amazing experiments that still have still ever been done in our field. She used ^{13}N as a tracer to analyze pathways of microbial nitrogen metabolism in soil. But ^{13}N is the radioactive isotope of N—it has a 10 *minute* half-life! You create it in a cyclotron[1] and then have less than an hour to finish the experiment before all the ^{13}N decays to ^{13}C. It was awesome work—those papers were why I applied to work with her. But then I saw this brilliant experimentalist who was no longer doing experiments. I didn't really understand. It took time for me to realize the truth of Maggie Rogers' distinction between *science* and the *science industry*.

Over time, I came to realize that as an academic and particularly as a professor in a research university, the job is not to do just more of the research that gets us here, but to move to the next level, to make research happen and to produce the next generation of scholars, teaching them to become awesome researchers and teachers in their own rights. Being an academic is more than just producing scholarship and teaching classes. Yet, most of us have to learn "the science business" on our own by trial and error. This is true across all areas of scholarship and academe—being an *academic* is different than just being a *scholar*.

We may have gotten into academia because we were fascinated by molecules, mealy bugs, math, movies, Marxism, or macroeconomics; but in reality, most of what we do is work with *people*. That becomes increasingly true as we move beyond graduate school to become professionals. I don't even know how to use most of the equipment in my lab anymore. But we rarely get trained in the "people skills" required to function in a human institution.

This is analogous to the famous maxim about military matters: "Amateurs talk about tactics, but professionals study logistics."[2] In academe, we see something similar—we read great papers, but we rarely think about the systems that trained those scholars, that allowed the work to get done, or to get written and published. Yet, once we move past graduate school, we come face to face with all those systems. They are fundamental to *being* an academic, whether you study soil biology or English literature.

I know of few books that talk about *being a scientist*, or about *being an academic*, rather than about how to do research, write papers, or teach classes. Ultimately, however, to be successful, productive, and happy as an academic means more than just doing clever studies, writing eloquent papers and books,

---

[1] By smashing protons into oxygen atoms (^{1}H + ^{16}O → ^{13}N + ^{4}He).
[2] Gen. Robert H. Barrow. Commandant of the U.S. Marine Corps. 1980. But such thoughts have been expressed by generals for millennia.

or even teaching great classes. It means succeeding in the professional arena that is defined by human and institutional structures—our departments, our universities, our societies, and our professional communities. Succeeding and thriving as an academic calls for developing wider, "non-academic," insights and skills into how these operate and how to operate effectively with, and within, them.

I.1 Academe is a Human Institution

Ultimately the organizations and structures that define academe are the human systems that dominate our lives and careers as academics. Functioning as an academic is about the relationships we develop with our communities of students, colleagues, and coworkers. Each group plays an important role; building a successful and fulfilling career calls for working effectively with each. Yet, because they play different roles in building our careers, the nature of the interactions differs.

Trainees: How do we teach and mentor people to get their best out of them and to help them launch their careers?
Campus colleagues: How do we work effectively with the people with whom we share a department, but with whom we might have few scholarly interests in common?
Academic Peers: How do we build effective professional relationships with our disciplinary colleagues at other institutions—i.e., the people who review and read our work?
Administrative and support staff: How do we work effectively with the people who allow our institutions (and hence us) to function?

I.2 Trainees: Students and Postdocs

Students are, of course, fundamental to our existence as faculty. So much so that most members of the public naturally assume that undergraduate teaching is why universities exist. Why else is it an inevitable question when we are introduced as professors: "What do you teach?" People understand our teaching mission. It is a critical mission, but teaching undergraduates is still only one aspect of what we do as college or university professors. Scholarship in all its aspects involves more than passing along existing knowledge; it also involves creating new knowledge, and importantly, new knowledge *creators*—undergraduate researchers and graduate students, some of who will become the next generation of professors. As teachers, we pass along existing knowledge; as mentors, we produce the next generation of scholars. Society relies on both—not surprisingly, so do our careers.

I.3 Campus Colleagues

We live in our immediate departments, but are surrounded by a wider circle of campus colleagues. Within our departments, most colleagues are in different sub-fields. Hence, only a few will be true professional peers. Mostly we interact with departmental colleagues in hallways and faculty meetings, where we deal with departmental management and politics. Some faculty members are content to live almost entirely within the bounds of their department, but that has become increasingly rare. Traditional hard-walled silos of knowledge have become porous. Neither nature nor knowledge can be easily and neatly packaged. As an ecologist, I live where biology meets the Earth sciences, so I connect with people in geography and Earth science as well as biology. Similarly, I know people who live in the fringes between environmental and political science, history and economics, and many others disciplines. Few of us are so centered in our niche that we have no connections to others; thus, our colleague base may span into several units, building networks of scholarly linkages that weave our campus together. But even with this, you will be lucky to have more than a few people on campus who read your papers, collaborate with you, or directly contribute to your scholarly endeavors.

I.4 Academic Peers

The best academic advice I have ever received was from my brother David, and it was simply "remember who your real peers are." I'd been kvetching about petty departmental politics, and about some frustrating colleagues. DavidHe focused my attention back on the people who ultimately build your career. For those of us in research universities, those are the people *outside* our home institution—our scholarly community made up of the people we have drinks with at conferences, who read our papers, invite us to visit for seminars, and who create professional opportunities. Importantly, these are also the people who write letters in support of our tenure and promotion cases. They may not directly vote on our cases, but departmental colleagues rely on their expert assessment to guide their votes—at a research university, the outside letters carry the case.

I.5 Administrative Teams

The last major human system that regulates our lives as academics is the staff team that keeps our systems working. Every university has more staff than faculty,[3] people who manage the systems that allow us to function. Our administrative systems may sometimes feel like they run on rules and policies, but they are truly run by networks of people. Whether those networks lubricate or obstruct operations varies from place to place, but we depend on them utterly to keep our institutions working. These people range from the custodians who keep our buildings working to the senior financial staff whothat manage the budgets to make sure the custodians get paid and keep the buildings working. Some of the staff we interact with closely—notably those who support us in our departments. Others are more remote. But they are all at the heart of maintaining a functioning university. Being successful means working within our systems and structures, and that means working effectively with the people who allow us to function.

This book is organized into four sections, each focusing on one aspect of the human systems that are fundamental to succeeding as an academic. The first starts in the center with *you* as you navigate through these networks to build your career. They span out from there toward the networks we engage with, from our academic department to our professional communities. Section 1 discusses the post-Ph.D. career trajectory as we move from graduate school and establish ourselves as professionals. Section 2 looks at universities in terms of the systems and people with whom we interact with. Section 3 focuses on our key training and educational roles—how to survive and excel in them. Finally, Section 4 looks at our wider human systems—the institutions outside of our university that are fundamental to our professional lives—publishers and our academic communities. I try to offer useful insight into how academia really works—and so how you can better navigate your way through it to build a happy and successful career.

---

[3] UCSB has ~800 ladder faculty and ~3800 permanent staff to support ~20,000 undergraduates and 3,000 graduate students.

SECTION 1
YOUR INDIVIDUAL PATH AS AN ACADEMIC

The Road goes ever on and on
Down from the door where it began.
Now far ahead the Road has gone,
And I must follow, if I can

                                                J. R. R. Tolkien

After we get our Ph.D. and launch our academic careers, our first focus is on ourselves: to get a job and to secure that job. How do we thrive as individual faculty members? The most important human systems are those closest to us: our research mentors and group members, and then our new colleagues in an academic department.

This section focuses on you as an individual faculty member. The first chapters follow the trajectory from Ph.D. through postdoctoral training into a position as an assistant professor and through tenure. Chapter 5 focuses on the extra challenges faced by academics who are not white heterosexual men. The last chapter addresses the issues faced by faculty who are not in tenure-track teaching positions and who actually comprise the majority of U. S. university faculty.

1
Postdoc

A Postdoc's Job Is to Get a Job

> *Beginnings are such delicate times*
> **Frank Herbert, *Dune***

> *Do you wish to rise? Begin by descending. You plan a tower that will pierce the clouds? Lay first the foundation of humility.*
> **Saint Augustine**

1.1 The New Ph.D.: "Green Ph.D." Syndrome

It feels amazing to complete your Ph.D. After a lifetime as a student, you aren't anymore—you're a *Doctor*. Wow! Getting a Ph.D. is a huge accomplishment, proving hard work, creativity, and intellect. You should be proud of yourself. I grew up in New York, though, where jokes on the subject usually have punchlines like "Pee-h-D, Schmee-h-D—why didn't you become a *real* doctor so you can cure my arthritis?"

In *Writing Science*, I wrote about the anticlimax of finishing my Ph.D.:

> After floating out of the library having filed my dissertation, I looked at the receipt they had given me and came crashing down. All my life I had been a student; my entire self-image was built around being a student. It's who and what I was. Yet that silly slip of paper said I wasn't one anymore. So who was I?

No longer at the pinnacle of the student world, I was now at the bottom of the professional. The most junior, inexperienced, and unproven Ph.D. in a world of Ph.D.'s. The degree I'd struggled for meant nothing; it was just an entry ticket to this new arena, where I had to start proving myself all over again.

Indeed, a Ph.D. is only an entry ticket that says you've finished basic training. Your mother might crow that "My Baby's a Doctor," but it won't impress anyone else. Your faculty all have Ph.D.'s, so it's no big deal to them. And the university staff who work with us? They know we can be just as clueless, just as careless, and just as inconsiderate as any "normal humans." They know that we don't just put our shoes on one at a time—they know we sometimes grab them from different pairs, or forget to put them on at all! Yet, we regularly hear of new doctors (Ph.D.'s and M.D.s) behaving badly, full of their meaningless new status. Get over it. Quickly. In fact, of course, when you've just filed your dissertation and been hooded as a Doctor, you *are* junior, inexperienced, and unproven.[1] There is still a lot to learn and to accomplish to become successful and build a career. Most of the key lessons, though, you are expected to figure out on your own. That is what this book is about—to lay out some of those lessons and provide guidance on how to navigate the pathways that will arise, and if not to build St. Augustine's "tower that will pierce the clouds," at least to get and keep a job.

Your first mission, of course, is to find a job that will enable you to become *less* junior and *less* inexperienced, and to begin to prove yourself and add some mass to your CV. Once upon a time, in some mythical past, that first, "trial" position might have been as an assistant professor. With the increase in applicant pools, though, it is hard to be competitive for a faculty position without some postdoctoral work. Hence, typically upon finishing a Ph.D., we look toward a temporary first position. This "Red Queen's Race"[2] has made it harder to compete straight out of a Ph.D. So we apply for fellowships, postdocs, or possibly lecturer positions to polish our skills, develop new ones, and build our record while getting paid (always a nice thing). By doing this, we position ourselves for a permanent position as a professor, researcher, or other professional in a university, institute, agency, or company.

In the natural sciences, postdocs have long been part of the academic structure—it's been decades since you could be competitive for a permanent position without having done at least one postdoc. Math departments hire "visiting assistant professors" as a postdoc-equivalent—you teach the core classes that are required for STEM majors while developing your scholarship with a mentor. In other disciplines, postdoctoral positions have been less common because there wasn't funding to support them. In the humanities and social sciences, you might once have been able to secure a faculty position straight out of your Ph.D. But as job searches routinely get one hundred

---

[1] This is comparable to medical training: You graduate medical school as an MD, but you're not ready to practice without an internship and residency (i.e., a "postdoc").

[2] From Lewis Carroll's *Through the Looking Glass*: "Now, here, you see, it takes all the running you can do, to keep in the same place."

or more applications, some kind of postdoc position has become increasingly common. Hence, for people interested in competing for faculty positions at research universities, some type of fellowship is increasingly routine. But aside from building up your CV, a postdoc is also a wonderful opportunity to spend time in a new place, make new connections, and do interesting new research.

1.2 Starting as a Postdoc: A Postdoc's Job Is to Get a Job

A postdoc position is temporary and defined by what you are *not*. You're *not* a doctoral student anymore—you're "post" that. But equally, you're not *anything* permanent yet. A postdoc is usually several years and is a mix of advanced training and polishing. The focus of a postdoc's professional life is to position for what comes next—a "real" job.

That seems pretty straightforward, perhaps, but beginnings are delicate times. If you are joining a new group or department, you arrive as a blank page and the first words you put on that page begin your story, a story that can be difficult to alter—except perhaps by tearing it up and moving, to begin a new story elsewhere. Joining a group as a postdoc can be a particularly delicate social dynamic because you may be a new "senior" person, but you're also the new kid on the block. You are likely joining the group to learn new approaches from people who may be your academic juniors, but equally by being more senior, you should be a mentor and leader who brings new ideas and perspectives into the group. Here, then, the St. Augustine quote about laying the foundation with humility is excellent advice.

Establishing your position in a new group isn't just about academic ability, it is also about finding your place in the social fabric. Group cultures may take their lead from the faculty leader, but they can evolve a bit like the *Lord of the Flies*, where the "kids" develop their own pecking orders and traditions.

I've been involved with many postdoc situations, beginning when I was a student, as a postdoc myself, and then as a professor. Most interactions have been fine, a few have been imperfect, and I've heard of horror stories. I've contributed to some of the imperfections. An early example was when a new postdoc joined Mary Firestone's lab. He joined several months before I finished my dissertation. Mary was on maternity leave and asked him to keep an eye on things, a natural and appropriate role for a postdoc. But I was Mary's first Ph.D., had been in the lab for five years, and had developed several of the approaches we used routinely (i.e., methods he was joining the lab to learn).

Equally, I was tired and stressed,[3] and a bit prickly about my status in the lab. He didn't recognize it—he was new. But my attitude was "You are only three months and three letters (P., H., and D.) ahead of me, and that doesn't mean squat." I didn't respond well to his stepping into the "senior" role. From my perspective, we started off a little rough, although we became, and remain, friends.

A year later, when I came back from Scotland, I moved to a second postdoc in the United States. Within the first few months back, I was invited to interview for two faculty jobs and also to present a lead talk at an international conference. My advisor was fine with it; he embraced the ideal that a postdoc's job is to get a job. But some of the other people in the lab seemed rubbed the wrong way: You just got here. Why are you away so much?

Social standing in a research group is based on several factors—talent and accomplishment, how long you've been in the lab, and, of course, personal charisma and leadership. My brother-in-law, Alex, was a postdoc in a large biomedical research lab where social standing seemed based less on actual productivity than on raw hours at the bench—how tough are you? Had he focused on that metric, though, I suspect he wouldn't be my brother-in-law! Thank you Alex.

Although joining a new group can be a delicate matter, most of the time things go fine, at least as long as issues are addressed openly. One situation that had potential for trouble was when Pete Homyak joined my group. Pete did his Ph.D. at the University of California Riverside and wanted to apply for a National Science Foundation fellowship, but to do so he needed to work with a mentor at a different university. He had an idea for a cool project, one I would be happy to be part of, but because of Pete's family situation, he couldn't move to Santa Barbara; he suggested that he'd come up regularly to meet and work in the lab but would otherwise stay in Riverside. I saw potential for breakdowns in communication and mentoring, conflicts over coordinating use of lab equipment, and so on. Would I be willing to work with him under those conditions? I took the gamble.

My lab members nicknamed Pete the "Ghostdoc," because most of the time he was invisible. That could have been a term of disparagement, but it actually reflected affection and respect. Pete made the terms of the arrangement clear, up front, so there were no surprises. He also brought his expertise in gas flux measurements to other projects we were working on, including one with a Ph.D. student who was visiting from Vienna, which led to a very nice

---

[3] I had a fellowship at the University of Aberdeen, but to take it I had to write about half my dissertation in under a month. I was pressured and stressed.

coauthored paper (Leitner et al. 2017). Despite his erratic presence, he was a great postdoc and has become a valued colleague and friend, but it could have gone very differently. Pete made it work.

1.3 Being a Postdoc

This is the easiest part—postdoctoral positions are all opportunities to focus on building your professional record. There are two main types of postdoc: the first is where a researcher gets a grant funded and recruits a postdoc to take lead on the work (this is common in the natural sciences). The alternative is where the postdoc gets a fellowship to support their research; they may still work with a faculty mentor, but they have greater independence; such fellowships exist across academe. Either way, there are few other formal responsibilities—you've proven you can do research, now do it! Your job is to produce scholarship and to write. In some fields, post-Ph.D. temporary positions may include teaching, such as with visiting assistant professors in mathematics.[4] But even in these situations, the rest of the time you're still supposed to be focusing on your scholarly portfolio.

The freedom to just do research is what most of us dreamed of when we started a Ph.D. program. New ideas, new insights, new places, new data? Research heaven! But the rub is, of course, that postdocs are temporary. You can't stay in this Wonderland bliss of pure research forever. We have to grow up sometime.

1.4 How to Stop Being a Postdoc

A postdoc's job is to stop being a postdoc—to get a job! Yet, there are numerous career paths open to people with Ph.D.'s. My trainees have ended up in careers as diverse as science policy at the Pew Center on Climate Science, the U.S. State Department, in a university library working to integrate researchers and library resources, in the Nature Conservancy, and even founding a company to produce microbial plant-growth enhancers, as well as becoming professors at both teaching- and research-focused institutions.

Hiring processes differ among jobs, but they share a common core. First, there is a paper application, in which you show your qualifications, why you are interested in the position, and why you think you are qualified. Phone or

---

[4] This works also because research in math is readily portable and rarely requires a lab.

video interviews may be used to reduce the number of qualified candidates to those who seem most suited, but the process will almost always finish with face-to-face interviews (in person or by videoconference).

Hiring is a fitting process, not a raw competition based purely on the fictional concept of "the best person." It's not about finding the "best person," but finding the "best person for you." When we hire an assistant professor, we are always looking to the scholar we hope you will become, not just who you are right now. That is why tenure exists—if it turns out we made a mistake, we can fix it. I have colleagues who say "just hire the best person," but I've had conversations with these same colleagues about the now-famous scholars we *didn't* hire. And of course, the decisions of search committees will affect the trajectories of our careers and how successful we ultimately become.

Recruitment is crystal-ball gazing—we can't know who will be the "best person" ten years from now. We recently hired a plant ecologist at the University of California Santa Barbara; almost all the candidates we interviewed had worked with Dr. Todd Dawson at UC Berkeley during their training. Todd has become an international leader in the field. Yet, when I was a new assistant professor at the University of Alaska Fairbanks (UAF) we interviewed Todd for a job—but we didn't hire him! Instead, we hired someone who doesn't now have the international stature that Todd does. Did we make a mistake? No. Not at all—we hired someone who was right for the program, a great contributor, and has become an established scientific leader in boreal forest ecology, which is an important, but somewhat constrained niche. Todd, in contrast, moved to Berkeley, where it's easy to recruit world-class students and postdocs who help generate new ideas, proposals, and papers, which in turn help recruit new great people, fueling the cycle. His record and reputation benefitted, driving a positive feedback loop; this effect on research productivity is profound, as I saw after moving from UAF to UC Santa Barbara.

When I describe recruiting as a "fitting process," however, I do not mean to imply that it is purely a popularity contest. It's not. To fit, you have to be damned good: We look for creativity, research productivity, and vision. But candidates often work on different topics and bring different combinations of skills and personal histories to an interview. what is most valued by any particular group will vary. For example, I applied for jobs in ecosystem science and competed against people who were plant ecologists: Which of us seemed "best" might have depended purely on whether faculty in the department leaned toward a plant ecologist or were open to a soil ecologist. I am friends with several of the people who interviewed for "my" position at UCSB. I was several years senior and had a different focus to my research;

I fit what the department was looking for at the time of the search. But am I *better*? Hell no.[5]

Of course, acknowledging that recruitment is about "fit" raises concerns about diversity—do we accept people as "fitting" when they are different than us? This can be a problem in terms of intellectual diversity, but even more so in terms of human diversity. Intellectually, the best person may seem to be someone who thinks like you—the same subniche of a field. Humanly, it had meant that all sorts of prejudices limited access, making it harder for women, Jews, people of color, LGBTQ+ candidates, and others to appear to be that best-fit, top candidate. Universities have made progress, some from shifting social expectations, some from campus initiatives (e.g., extra funding if you hire someone who advances diversity goals), and some as a result of beginning to expand diversity in our departments and among our trainees. As programs diversify, the barriers lower for the next generation of hires.[6] But if you are currently looking for a job, there's not much you can do about programs to which you are applying. Just recognize that all of what goes into defining the "best candidate" might be hard to predict and is out of your control. All you can do is apply and see—is this a place that wants me and that I want to be part of?

If you get hired into a position where you are not a good fit, you won't stay. You'll either get frustrated and leave, or you'll fail and get fired (denied tenure); the alternative would be to be chronically unhappy. Those are all miserable outcomes. You need to work to find the right fit. Part of that is knowing who you are, what you are willing to do, and even where you are willing to live. Don't apply for jobs that don't suit you, and make sure that you present the "real" you. Job hunting is like e-dating, where the application is your profile. You can touch up your photo and lie about your interests or past. That might get the search committee to invite you for a "date"—an interview. But what do you do when they figure out that you don't really like moonlit strolls on a tropical beach and the opera—rather, you'd prefer to be hiking in the Scottish highlands and listening to bagpipes? You don't need to describe all your warts and weaknesses; rather, present your best self—dress nicely, pull out your best stories, don't get drunk at dinner, and so on. But it still needs to be you, so that you can figure out whether this is a potential match.

I interviewed for a job at Kansas State University when I was first looking. For a soil biologist, it was a great position; K-State has a world-class program.

---

[5] That isn't false modesty. I may be a "Highly Cited Researcher," but one of those "competitors" has been elected to the U. S. National Academy of Sciences.

[6] We have developed tools to help reduce barriers, including screening rubrics to establish qualified candidates, external oversight of candidate lists to ensure that qualified candidates are not left off, and so on.

But I grew up in the other Manhattan—the one in New York. *The Big Apple.* Far from "The Little Apple" of Manhattan, Kansas, in both miles and mindset. I gave a strong seminar and had good meetings with faculty, but I heard later that I put off the dean and the department chair. I've always guessed that was because when they asked about the short postdoc I had done in Scotland, I said that I didn't do it just because it would advance my career, but also because I thought it would be a lot of fun. That was true—I had a blast. But, it also allowed me to expand my skill set and connect with ecologists in Britain and Europe, a path along which I have continued. But if it showed the dean that I was not the right person for them it was the right answer.

Shortly after that, I interviewed at the University of Alaska Fairbanks. I almost hadn't applied—Alaska? Really? I applied because I needed a job, and the job description sounded like a good intellectual fit; I figured that if I got an interview, it would be good practice. My response to my mother's semi-serious comment that "You're not allowed to move to Alaska" was "don't worry mom, I'm not." But then I arrived in Fairbanks, and Dr. Terry Chapin picked me up at the airport. Almost his first words were "You must be tired, so I can take you back to the hotel for a nap before dinner, or if you'd prefer, we can go skiing." A famous ecologist is going to take me cross-country skiing on a *job interview*? That defined UAF for me—do good science, but have fun doing it and appreciate the environment you live in. Why else live in Alaska? It was love at first sight; I was desperate to get an offer before we got back for dinner.

Some of my students have run into problems with the fit process in job searches. One did his Ph.D. on methane cycling in Alaska and was looking for jobs when geomicrobiology was emerging as a field. He wasn't really a geomicrobiologist, but he fit what some job ads appeared to be looking for, so he applied for those. At one university, a faculty member pretty much said to his face "You're not a geomicrobiologist—why are you here?" My student was flummoxed and responded with "uh . . . because you invited me?" He'd been honest about who he was in his application, but the search committee hadn't considered what they really wanted and invited him because his record was appealing; as a result they wasted time and money and treated him badly.

This raises the question of what you can do to figure out what a program is looking for and how you can make sure you show them the aspects of your true self that *will* work for them. You might like dressing up and going to the symphony, and you might equally like putting on cowboy boots and going to a honky-tonk. But, which is best to invite someone to for a first date? Sometimes, there is no way to really know, as with my student's geomicrobiology fiasco. He fit the job ad, and he was honest in his application. But the department essentially lied about loving the symphony! They were impressed enough by his record that they invited him for a date even though it should have been clear that the date was going to be a failure.

In contrast, I had the opposite experience with UCSB. The job ad I applied to was for an ecosystem ecologist, but I considered myself a microbial ecologist, and I knew who else was interviewing—I didn't expect an offer. I didn't know the department's history and perspectives, and that my microbial focus would be seen as a positive; I'd figured it could well be a negative. I had no way to know, and as it turned out, I fit well when I did not expect to.

Many of us may have made errors of judgment in our personal lives, being drawn to an attractive body or entrancing smile, while overlooking deeper signals of incompatibility. In academe, a strong CV is attractive and can draw invitations, some of which are doomed.[7] Interviewing is time consuming and emotionally draining. And what would you do if one of those wrong partners asked you to "marry"—that is, they offered you a job. Do you take it because, well, it's a job? Or do you turn it down knowing that there may not be another offer out there? There are no easy answers to these questions. Sorry. But recognize the challenges and that the key is to know yourself, and to be yourself. Everything after that you'll have to play by ear. In my career, the decisions that have worked out the best are those where it felt "right," rather than because of any elaborate calculations working through all the pros and cons.

1.5 The Job Interview

When departments run job searches, we really all have one question: "*What can you do for me?*" In a faculty search, that translates to some mix of questions such as: "Will you teach classes that we need covered?" "What research tools will you bring that I can use?" "What proposals can we write together?" "Will you support my political agendas in the department?" Will you enhance the department's reputation or ranking?" For some people (those intellectually further from you), it may just mean "Will you take the job so we can move ahead with the next search—the one I actually care about?"

It is universal advice to do your homework—check who you are going to meet and read some of their work to get a sense of what they do and how you might answer their version of "what can you do for me?" On the other hand, knowing too much can make the process awkward. I interviewed for a job in a department where I had several friends who wanted to hire me. But, there

---

[7] But don't forget my Alaska experience: Before I flew to Fairbanks I was confident I didn't want their job. They showed me otherwise.

were political divisions and other factions had different ideas. My friends gave me *lots* of detailed advice on how to handle conversations with different people—so much it made the interview awkward, and they didn't offer me the position anyhow. You don't want to seem clueless—that can suggest you don't care. But how do you be yourself if you know so much that you are parsing each conversation trying to be what you think they want you to be?

1.6 The Seminar

The core of an interview for a faculty position is always the job talk. It's where you strut your stuff and show us how cool your scholarship is. This is also where the faculty get a perspective on whether you are likely to be a good teacher.[8] It's all-important. I've seen candidacies fail in the opening seconds of the seminar. In one, the lights dimmed, the title slide came up, and ... the candidate turned sideways so that both screen and audience were visible in their peripheral vision, and then started speaking to the wall, so quietly that it was almost mumbling. Few people may crash and burn before they even get off the runway that badly, but a weak seminar is the number one reason that applicants fail in a job interview.

The key in a job talk is to remember that the department is interviewing you because they don't already have *you*. They lack strength in your area and are looking to fill a niche. But that means that most attendees are likely not expert in your area of scholarship. If you are interviewing in an interdisciplinary program, they may not even understand your communication approach.

For example, in the natural sciences, reading a paper at a job talk is anathema—grounds for immediate disqualification. Yet, this is common in the humanities. For years, I wondered about this—if you're just going to read the damned thing, why not just give it to me and let me read it? But one day I had a realization. In the sciences, we take great care in how we present our most important material: our data. We are careful in how we plot data, prepare graphs, and structure diagrams. The exposition in between is less vital, so we're more relaxed with that. My realization was that humanists are the same, except their most important material is concepts and ideas. They don't present these in a graph, but in a carefully assembled piece of language. To ensure that language is presented precisely, they write it out and read it. The best talks I've seen in the humanities do exactly what we do in the sciences: open,

---

[8] For some positions there may be a second "teaching talk" to see how you present material to a student audience.

casual exposition for much of the talk, but then going to the prepared language to nail the key points. I've served on search committees for positions as divergent as environmental ethics and plant physiology; good talks all achieve the same goal—they educate a wide audience about the material. A good talk weaves an overall story but still shows your deep scholarship at key points.

My rule is that everyone in the room must understand why your scholarship is important and what you have contributed. That means the opening must be broad enough to engage people who are not expert in your subject area, and it must define the problem in terms they can follow. The conclusions, equally, must speak to the entire audience, showing what we have learned. If you target too narrowly in these key sections, you're toast. But if you stay at a general level for the entire talk, you are, equally likely, toast. We want to see your deep scholarship, even if we don't follow all the nuance. There must be an overall story arc, but each section should have its own internal arc, where you can introduce a piece of work, show us how it fits into the whole, and blow us away with the sophistication of the analysis. We might not follow the detail of the molecular biology, econometric model, or ethical nuance, but we can see it's there and how you use it to develop the larger story. We want to see that you can do the most sophisticated scholarship that will impress your disciplinary peers, but we also want you to teach us and to show us why we should care about your work. Make sure, however, that when you take us on those deep dives into detail you don't stay very long, because we'll drown. Show us your substance in small pieces, then come back out to show us what it means, preparing for the next arc and the next substance.

Job interviews can be fun, as long as you don't get too wound up in it. Have fun on the date, who knows—maybe it is true and lasting love?

2

Assistant Professor

Making It to Tenure

An assistant professor's job is to keep their job.

2.1 Starting as a Professor: From Offer to Opening Day

Beginning as a new faculty member is a delicate time: You are establishing yourself in an institution where you might spend the rest of your career. Your new department has just run a national or even international search, evaluating possibly hundreds of applicants, to find the person they think will be best for them, and they found *you*. Wow!

Your new colleagues are going to be excited about your arrival, but they don't know you yet. They met you during the interview and a few hosted you to dinner. Although most probably attended your seminar, some only heard what the search committee reported. You must have made a good enough impression that they hired you, but interviewing is like dating. Once you move into your new office, though, you're living together—now they get to see you before you've had your morning coffee. Don't give them buyer's remorse!

2.2 Before You Sign on the Bottom Line

What do you do when you get the phone call offering you a job—at least after you stop jumping up and down and shrieking with joy? You need to think about start-up negotiations. What is negotiable? With some jobs, you'll only have one question: Take it or leave it? For other positions, there may be a lot to sort out: start date, salary, space, and start-up expenses are all potentially on the list. But how you approach the process may influence your life at your new institution as much as what you actually negotiate. How you handle the transition from candidate to colleague sets your new department-mates' impressions of you. Negotiations are important in that. We expect you to negotiate the best arrangement you can, but if you handle this poorly, you might start Day 1 with colleagues thinking you're a jerk and that they might have made a mistake. *Not bright.* Negotiating calls for balance. You ought to push on the issues that are possible for you to push on, but in a way that the department sees that what you are asking for is reasonable and may even benefit them. Also, know when to stop pushing.

The culture around negotiations differs among fields, institutions, and nations. In the United States, there is almost always room to negotiate on salary, and new faculty can expect a budget to allow them to set up and function. How much is allowed varies, of course, because where a natural scientist needs a lab, a mathematician or a humanist may only need an office. But some elements of negotiations are common across the various fields.

One question that should help define your negotiating approach is, "Who are you negotiating with?" *Whose money are you asking for?* Are you talking to the person who controls the budget? Or are you talking to an intermediary? For example, at the University of California Santa Barbara (UCSB), new faculty negotiate via the department chair. But start-up money comes from the dean. Thus, the chair is your ally and advocate—getting you a good package is a way to support you, to launch your career, and to build your loyalty to the department.

The chair won't want to annoy the dean by asking the impossible, or by pushing so hard for you that it might piss the dean off and create problems for the future, but a chair succeeds by pushing for their people. When I was chair of Environmental Studies, the most fun thing I ever did was tell someone that her start-up request was unreasonable and that I wouldn't even consider it as a starting point—I wanted to see a request twice that size. Within the norms of her area of the social sciences, her request was reasonable, but I reported to the dean of sciences, who was used to million-dollar packages. Anything she got was going to be the dean's pocket change.

At other institutions, however, you may be negotiating with an institute director or directly with a dean: the person who controls the budget. It's *their* money and they'd like to spend as little as they practically can on you. They want you to accept the offer, and they want you to succeed. They recognize that means getting you a solid package, but they know what else they could do with that money, and so might be less inclined toward generosity. In UCSB's model, if the chair says "don't ask for that," she is probably just being honest about it not being possible; if an institute director says "don't ask for that," ask her "why not?"

It is also important to identify what is reasonable for the institution. In chemistry, at a top-tier research university, it may be reasonable to ask for a package that runs $1 million or more; but at a smaller institution, such a request might generate laughter, or worse. In the humanities, a new faculty member might only get new office furniture. The department chair, and other faculty, will likely guide you in the process. What is negotiable? What are the limits? You can try to push the boundaries, but it's unlikely that the chair is out-and-out lying to you. If she says "we have around $300 K for start-up," you might try asking for $325 K. You might aim for $350 K or even $400 K if you have a compelling story for how the extra would create a new resource for the department, but I'd expect the argument to fail. And that million-dollar instrument? Dream on.

One of the cleverer approaches I've seen to push up a request was when UCSB established the Bren School of Environmental Science & Management. Bren was recruiting a microbiologist, who pointed out that in any similar position, core hardware such as autoclaves and centrifuges would exist in a shared-use facility and were essential to teach a lab class. Bren should establish a core facility, which should not be part of the new faculty member's individual start-up package. The argument worked—largely because it was *true*. If you are going to push boundaries, offer arguments the chair can bring to the dean, and could potentially use to explain to the next new hire why you, apparently, got a bigger package than they did.

Of course, some issues aren't about cash. For example, teaching expectations: How many and which classes will you teach? Do you get reduced teaching for some period? How much? How long? Such issues are likely under the control of the department chair. The official offer and commitment letter, however, probably will come from the dean and might well not address such departmental issues—so make sure you get a letter from the chair on such issues. Otherwise, they might forget, or might rotate out; the new chair might not know about any such agreements and might not feel obligated to honor them. Bottom line—for anything that is important to you, get a clear written record that confirms the agreement.

The most challenging request you can make is for a job for your partner. The grad school years are when we often partner-up. Somewhere back in the "old days,"[1] almost all faculty were men, the only socially acceptable relationships were heterosexual, and you could support a family on one salary. Thus, partner-hires were a nonissue. None of those conditions, however, remain true. We often meet our partners in our classes, in the library, in our labs, or at our field stations,[2] and both partners need jobs. But hiring faculty is a

---

[1] My vote would be to call those the "Bad Old Days" as they relate to gender issues in academe.
[2] Close proximity with zero pretense: being grubby from days in mosquito-laden drizzle strips you down to your essence. There's no putting on your "date-face."

huge, long-term, commitment for a university—salary, office space, teaching assistants for the classes they will teach, and so on. Most universities hire good people and set them up to succeed, so the partner's position has to be fully supported.[3] Creating a new position—one that wasn't in the plan—is *never* trivial. It starts with departmental politics: How will this influence recruiting in other areas? If the partner needs a position in a different department, the discussions take on an extra dimension—why should Sociology create a position to help Biology? The obvious answer is "what goes around, comes around"—next year Sociology may need the favor. At UCSB such favors are facilitated by the first department "paying" for the new position—it's credited against them. But few are so naïve as to believe that means the "free" position is *free*: there's no such thing as a free lunch. Or a *free faculty line*. Later, when the university tallies up whether a department has sufficient faculty to cover programmatic needs, that "free" position will count in the department where they teach classes.

After departmental politics are resolved, the higher levels of administration get involved: deans, vice chancellors, and academic senate committees. The department justified *your* position by defining a compelling need in their teaching and research agendas. But now to recruit you, the university must upend its academic and financial planning to fit your partner into the mix. Universities have gotten better at this—we recognize that academics often come in pairs.

And don't forget that once we hire *both* of you, we have *each* of you. There is no guarantee you will stay together, but we have made commitments to each of you individually—tenure is tenure, and it is based on each person's individual academic accomplishment and trajectory.[4]

Creating new faculty positions is difficult for most universities, but it is at least straightforward. What if the partner is a bit more off track? For example, a partner who's director of a field station, something that might not exist on your campus, or if it does, already has a director! With such needs, negotiations are likely to take time, creativity, and flexibility: We might not be able to offer perfect positions for both partners, so what can you accept? If the university makes vague promises of something in the future, should you trust them? Once you accept an offer, you lose your negotiating leverage, so can you rely on those promises? Get things in writing.

---

[3] The additional position may be a little cheaper—for example, just one housing package—but only a little.
[4] I know a couple who broke up in between accepting positions and actually starting—good thing their University was happy to hire both!

The other tricky issue with asking for a partner hire is when to raise the issue? During the interview? That might reduce the likelihood that people would support you, even though deciding whether to offer you a job based on who you are partnered with raises moral and legal issues. But motivations can be masked—they could always come up with a valid-sounding rationale for offering the position to someone else. Alternatively, if the department never gets a whiff of the issue until they've offered you the job, they might be not be pleased to be blindsided. What if there is simply no way they would be able to create a second position? It would be easy to end up in a situation where no one is happy—a failed search for the department and the possibility of imperfect gossip in your field about you.

Of course, if you and your partner are in the same field, people in the department probably already know your situation and will likely bring it up in conversations. It might be inappropriate for the search committee or department chair to ask about your partner, but I'd expect the issue to arise informally in conversation; possibly from someone who knows you and is figuring out how to muster support for your request! There is no single, or even good, template here—just be careful feeling out what to ask and when to ask it. I wish I had a clearer answer, but every situation will be different, and will require sorting out.

Your start-up package defines the resources you begin with, and at many universities salary raises are by percentages, so your starting salary can regulate your income for many years—perhaps your entire career. Negotiating well is important. I made mistakes at both the University of Alaska (UAF) and at UCSB, but those at UAF threatened my career. I was so excited when they called and offered me the position, I just said "yes," including to the lab they'd shown me. I saw it was small, but I hadn't thought how quickly 20' x 20' fills up when you start putting stuff on benches. A few years down the line, though, I'd developed a side project with a postdoctoral researcher in a different group. We were borrowing space in her professor's lab for the work, and we wanted to scale the project up to submit a full proposal to the National Science Foundation. To do that, I needed ~8 feet of committed bench space for working with radioactive ^{14}C—space I didn't have. I went to my institute director and asked him to identify the space, should the proposal be funded. He said "I'll pass your request to the space committee." That would have been fine if we'd had a functioning space committee—but we didn't! I was happy at UAF, but if I couldn't get support for my core needs—to write proposals, do science, and publish papers—what were my options? That sparked a career-threatening political crisis that included my saying impolitic things to the new chancellor of the university and leading a drive to push the director to retire; things you should *not* do as an assistant professor.[5]

There are several morals to this story. The important one here is that because I allowed excitement at a job offer to overcome critical analysis, I negotiated poorly and so set myself up to fail. Either I wouldn't be able to do my work or I'd get caught in a potentially fatal political mess—likely both. Had I more carefully evaluated my needs, I probably would have seen that I would outgrow the proposed lab space. That would have allowed me to either start with better space, or at least to get a written agreement that should my space needs grow, the institute would find the needed space.[6]

2.3 Beginning

Starting a new position is delicate, because first impressions are strong and hard to alter. Missteps can be damaging. The most extreme case of "fouling the nest" I've ever seen was actually a senior hire. This person negotiated a year off to wrap up commitments, but during that year, they made enough demands of the office staff and the chair to be not only a nuisance, but a serious irritant. A few months before they were supposed to start, they called to say they'd decided not to come after all! At that point, even those who had been most excited about the hire joined the celebration at having dodged a bullet—they had realized that offering this person the job had been a mistake and were dreading their arrival. Try not to squelch your new colleagues' excitement for at least a year or two.

Shortly after I joined UCSB, I had a serious fright. The National Center for Ecological Analysis and Synthesis (NCEAS) had just been funded by the National Science Foundation and was searching for a director; I had a friend on the search committee. I had a few questions about the first candidate's research seminar, which was in my friend's area. That started an e-mail conversation, in which I commented on several of the candidates. When a senior departmental colleague found out, he reamed me out severely for "lobbying" the search committee. *Oh shit!* I was horrified that my new colleagues might get the impression that I played backchannel politics to manipulate decisions.

---

[5] I did get a bigger lab, but left a few years later to be with my wife, Gwen.
[6] That might not ultimately have avoided the problem, but it would, at least, have given me a better starting position.

I had opinions about the candidates, but it hadn't occurred to me that I was "lobbying" my friend or that I was sticking my oar into a delicate political situation. I was new and I was worried, even terrified, about making a bad initial impression with an influential colleague. Thankfully it blew over—in part because the decision of who to hire became clear.[7]

2.4 Establishing Yourself in Your Department and on Campus

A department is like a marriage—we partner in youth and grow old together. Nine people in my department have been here since I arrived—we've been sitting in faculty meetings together for over a quarter of a century. We've learned one another's foibles, developed friends and allies, and now we're growing old and crotchety together. Academic politics thus become a hybrid of institution and family—debates over vision, direction, and resources combined with petty family squabbling. The participants might have even forgotten what started the squabble twenty years ago.

Establishing yourself in a department—building respect, credibility, and influence—takes time as your career and relationships develop. We may start out full of energy to do cool scholarship and transform our communities, but it's wise not to annoy people who vote on your tenure case. Senior faculty can get frustrated with colleagues who join a department with the energy (and tact) of a puppy, barking a lot and insisting that you throw their ball, *now*. I've had colleagues who, from their first faculty meeting, were vocal and could rub me a little raw, strongly expressing views even on topics with which they didn't have experience. But, we hire people to enhance the department and to create its future, and we rarely hire people who are shy and retiring. If they'll put up with my mouthing off, I'll put up with theirs—I'm sensitive to the "pot–kettle issues."[8] Some departments, however, are personal and political minefields that are not safe to navigate without the armor plating of tenure. Normal advice would be to chill and figure out how systems work before becoming politically vocal within the department—I don't expect, however, that many will take that advice!

---

[7] Dr. Jim Reichman was a great director.
[8] From the parable about the pot calling the kettle black—accusing someone of something you are equally guilty of.

Remember that processes can play out over years; accomplishing *anything* in academe calls for patience. Consider recruiting a new colleague: It can take several years to get the position to the top of your department's wish list, several more before the position is approved, and most of a year to run the search. The person to whom you offer the job might then put off coming for another year. Or, the search could fail and have to be repeated. Hiring that new colleague might take the better part of a decade.

Stature, personal and political, in a department generally correlates with your external professional reputation, although the correlation is weak and has many outliers. Being a professional big-wig who brings in recognition and grant money usually translates into status in your department, but our perception of our colleagues' professional standing is often imperfect. Departmental colleagues are usually in different subfields and mostly know our records from reading our merit and promotion cases, or from hearsay, rather than from reading our papers. But they still see and evaluate how we operate within the department.

Figuring out how to operate within your program is something you will have to sort out and to which you'll have to develop your own approaches. The easiest and fastest (as well as more pleasant for all) way to build stature is through community vision, building effective relationships, and supporting your colleagues. Even in academe, though, some people take a different path, building their own little empires, aggrandizing personal power and glory. That approach can succeed if the empire builder is a skillful enough politician and has the scholarly stature to translate into political clout. Of course, this is a high-risk and lonely strategy; even if you succeed, you will likely be disliked, and if it fails . . . well, after Waterloo comes exile.

Working effectively with colleagues doesn't mean you never butt heads. I have a colleague I revere, but we often seem to be on opposite sides of issues in faculty meetings. We both recognize that we are always on the same team, working for the good of the program, but we just see things differently. We work well together and I value their style of "let's put all the cards on the table and see what we can assemble." We've done well through that approach, and our relationship, I hope, fits within Dr. Emily Berhardt's wisdom in her column, *Being Kind* (Bernhardt 2016): " . . . we hone our understanding through challenging our own views and the views of others . . . Yet we can spar while still being kind. We can disagree with a point while respecting the person making it." That sums up how I feel about this colleague, although I sometimes worry that new faculty might misunderstand our interactions, which can sometimes appear combative.

2.5 Developing your Support Network

When we're new in a position, we have many needs in order to establish ourselves and succeed. We look for role models and mentors, but it's unlikely that we will find one Yoda-like guru who can answer all our questions and meet all our needs. Such magical mentors don't exist. The National Center for Faculty Development and Diversity (NFCDD) identifies a list of support needs that new academics are likely to have and may look for a mentor to help with (Table 2.1).

No one human is likely to meet all these needs. The person best able to provide emotional support, for example, might be someone who was hired around the same time as you; they might, therefore, not serve as a role model or sponsor. Your best role model for academic life might not be in your research area, and so might not be able to sponsor you to open doors to new opportunities. That's fine. Different people can serve each of these support and mentorship needs—they may not be in your department or even on your campus. You need to assemble your own personal and professional support network as you figure out what works for you.

2.6 Professional Communities

Chances are that in your department only a few people read your work and truly understand your scholarship. After all, they hired you because

Table 2.1. Types of Support that New Academics Need

Professional Development	How to do all the things we do as academics? We need to learn the professional skills. *I assume that is why you're reading this book.*
Emotional Support	Who celebrates our successes and commiserates with us over our struggles?
Intellectual Community	Where are the people who are our scholarly peers? Maybe other departments with overlapping intellectual space or faculty and researchers at other institutions.
Role Models	Who are the people we admire and try to model ourselves after? Or perhaps parts of ourselves?
Access to Opportunities	People who know the systems and can help connect us to opportunities.
Sponsorship	People who use their influence on our behalf and "shape the story" of who we are, and who support us into new roles.
Substantive Feedback	How am I *really* doing? What could I do differently or better? Who can we trust to be fully honest?

they had a gap that it's your job to fill. It's possible that you were recruited to add strength to a cluster of excellence and will find yourself surrounded by close collaborators; if so, you are a lucky exception. The department may be your home base and your family, but they are likely not your professional colleagues.

You may find moral and intellectual support in your professional community—the people who work on the same topics as you do but who are spread across the nation and across the globe. They form one of the essential human systems we rely on as academics; important enough that Section 4 of this book is about that larger "external" ecosystem. The connections we develop with our larger peer networks will ultimately be as important as those within our department. The people we hang out with at conferences will likely be our friends for our entire career. That group is, however, likely open and fluid; my goal for every conference I go to is to add one new person to my circle of "friends and relations." Over time, that adds up to a pretty large circle. The relationships you develop early will stay with you for the rest of your career. They are valuable and precious—they will sustain and support you.

2.7 Time Flies When You're Having Fun

Six years goes by quickly. The time between sitting down in your new office and putting in your tenure package will feel like a blink. There really isn't time to settle in and catch your breath; if you do, you might look around about a week later and realize you're already sliding behind the curve. Aim for a fast launch and have a plan (or at least a vision) of what you want to accomplish. We work differently across the communities of academe, but we all work hard, beginning from Day One.

In areas of scholarship where we work in research groups (the natural sciences and some areas of the social sciences) there is a "momentum cycle": Grant money allows you to hire people (students, postdocs, technicians) who generate data and write papers, which motivate ideas, which stimulate new proposals, and which bring in money. Once the cycle is spinning, your people provide some of the energy to maintain it. But, *all* the energy to spin the cycle up comes from you! And we expect you to do so. In areas dominated by individual scholarship (humanities, some of the social sciences, and even math) though, it's different—when your fingers aren't on your keyboard, activity ceases. In a research university, there are key steps to get rolling, and they involve some measure of luck—getting grants funded or finding a publisher for a book. In teaching-focused positions there may be less luck, but no less work—you need to develop classes and they must be good. Regardless of your area of scholarship, this all has to happen while you are engaging with your department and establishing your life in this new place.

Succeeding as an assistant professor calls for staying focused on the essential activities. Do you have a plan for how to prioritize and achieve your goals? Remember, however, Murphy's Law and its corollary that "no plan survives contact with reality."[9] Allow flexibility to take advantage of opportunities, to build new collaborations, and to deal with the glitches of things going wrong. But stay focused on the target—your scholarly and teaching records. Everything else is ultimately secondary. Be selective and take on tasks because they advance your professional goals and your ultimate mission: *keep your job*.

2.8 Learn to Say "No"

Scholarship is about knowledge: We learn early that to succeed in academia we have to *know* a lot. We're taught less that we also should *no* a lot.

Learning to say "no" is important, but it is a hard skill to develop. I'm still terrible. I have often said I want a poster for my office wall with an international "no" sign superimposed over a big "K" to remind myself that when asked to do stuff, the default answer should start at "no"—the version without the "K" (Figure 2.1). There are only twenty-four hours in a day and seven days in a week. We must prioritize. Yet, despite often saying I want that poster, what is actually on my office wall is evidence of having done the opposite—certificates related to having said "yes" to stuff.

That shouldn't be a surprise. Our departments and universities have work that must be done. So do our professional communities—and particularly early in our careers, being asked to serve on committees, panels, and editorial boards is an honor and a sign of professional recognition. "They want me to do that? Wow. They must think well of me!" So "yes" is a natural inclination—you don't get hired if you are unwilling to put yourself forward. But don't forget: As an assistant professor, *your job is to keep your job*. First, foremost, and always. That means building a strong record in your primary missions of research and teaching. Service is also a core part of the job—you are expected to do appropriate university service—and really strong service may get you a bigger

---

[9] The original version comes from the military: "no plan survives contact with the enemy," but it seems universal.

Figure 2.1 "No" is not spelled with a K

raise[10]—but it *never* replaces achievement in the critical areas of scholarship and teaching when it comes to tenure or promotion. We don't grant tenure to people who are great committee members but don't publish.

Professional service does support a tenure case by showing that you are on a successful trajectory. The National Science Foundation or the National Institutes of Health invited you to serve on a panel? A major publisher asked you to review book proposals? These show that the people who matter think that you matter. Such activities indicate success. But they can, if overdone, undermine that success. If you do well in one thing, people will ask you to do more. It is easy to get loaded down with these projects, which although valuable, take time you probably could better be spending on scholarship. Service can become a form of "virtuous procrastination" (Hayot 2014). Protect yourself—learn to say "no."

Particularly if you are a woman or a person of color, universities will ply you with requests for service—jobs that are essential to the university and to our goal of creating a more diverse institution. Yet, agreeing to all those requests can reduce your scholarly productivity. Until our institutions can create a more diverse faculty, this Catch-22 will persist: If we don't have diverse representation in search committees, resource committees, and so on, we are behaving badly. If, in trying to protect these faculty, we don't invite them to serve in these roles, we are being patronizing—protecting them from themselves, but on *our* terms, and so, again, behaving badly. If we achieve

---

[10] At UC, for example, we have an step system that determines salary and raises, and you can accelerate steps for service, particularly in leadership roles.

diversity by overloading a few critical people, we may be setting them up to fail. Knowingly doing that is worse—it's evil. But we don't know your limits or interests.

So we will invite, but if you are one of these "over-asked" people, hear in your own head: "We're inviting you to serve on this committee, but my advice would be to politely tell us to take a hike."[11] We need to invite you, to serve institutional and societal needs, but even more, to meet those same institutional and societal needs, we need you to succeed!

Those objectives are in tension with each other, setting our institutions' short-term needs potentially at odds with long-term ones. Don't fall into the trap—choose your battles. Even simple battles can be stunningly time consuming. Remember that your ability to create change grows with your professional rank and stature, and that you likely spend only six years as an assistant professor in a career that might last forty years or more.

You can be over-asked even if you aren't a member of a minoritized group—there is always a group of "usual suspects" who show up in academic senate committees because they are drawn to leadership roles—building the university for others. Campus leadership is one career path that some faculty embrace following tenure or promotion to full professor. But equally, I know faculty whose single greatest service is that they do none.

Being committed and good in the jobs you take on is important, so only agree to those that you care about and want to put energy into. It's OK to say "no" to the other requests and to make that "no" stick. If people keep pushing, push back: "You want me to succeed (finish my book, get that proposal done, serve my students), don't you?" Or, perhaps offer an alternate approach: "I don't have time to serve on the committee, but could consult occasionally or comment on documents you produce." Or negotiate: "You really need me to serve on that committee? I can do it if I don't teach introductory European History next year."

2.9 Leaving: Breaking Up Is Hard to Do

Most of us accept the first faculty job we are offered. Some have had the luxury (or crisis) of having multiple offers, but it's rare. When we're starting out, we apply for every job that moves, and we don't necessarily accept a position because it is a love-at-first-sight, passionate, lifelong romance. Many of us

---

[11] Or impolitely—just tell us how to fold it up and where to stuff it! But along with that, have some sympathy when not every committee is as diverse as it should be.

"settle," figuring a job is a job is a job, and any job is better than unemployment. Even if we have other, possibly "better" applications in the works, we can't risk saying "no" in hopes that some better offer materializes—not when jobs might have hundreds of applications and when we consider the competition, that is, the other enormously accomplished and talented people who applied. We also knew, when we started down this path, that we might not be able to pick where we end up living.

As an untenured assistant professor, you are in a trial period—the university hasn't sworn lifelong loyalty to you. You, equally, have no obligation to commit your lifelong loyalty to the university! Your first obligation is always to yourself and your family. You did, however, accept a job, and in doing so not only blocked the department from hiring someone else, but also cost them time, energy, and potentially a boatload of money. Please don't treat a first job as just a casual fling—this is a serious relationship to which you owe commitment and loyalty, at least until you realize that it isn't working for you.

When is it time to break up? What might lead you to start looking around for a new relationship? There are several common themes:

1. Frustration with your current institution
2. Professional opportunity
3. Personal relationships

Some people get frustrated with how their department or university functions and so start looking for a new position; I almost did when I was at the University of Alaska. Keep in mind, though, before you decide to divorce because you're having a spat with your partner, that any new partner will have their own quirks and imperfections! If you're considering leaving because you're frustrated, my advice is that unless you have become seriously, even toxically, allergic to the specific problem, live with it. All our institutions have glitchy systems and people who can be annoying. Chances are that if you move because you're frustrated, you'll just find new and different things with which to be frustrated.

One way to address departmental problems can be to shift to a different department at the same university. This is tricky, however, even if there is another department that is a reasonably good fit for you. Telling colleagues in your department that you want to leave them can cause bad blood. That is, however, probably the reason you want to leave, so maybe you don't care. But the new department will pay a price—they will be getting a new body and new teaching power; for them, it's equivalent to recruiting a new faculty member. But they already have you as a research colleague and their students can take your classes. So what is the benefit to them for taking you on? And if you don't get along with your departmental colleagues, will the new department think the problem is those colleagues—or you?

Should you get that grumpy about your current department, start with a delicate conversation with a friend in the potential new unit to get a sense of the politics associated with moving: Is it possible? Then talk to the dean who would have to approve the transfer. Will she think that if she doesn't support the move, you might instead just leave campus? Are you valuable enough that she would see losing you as a serious blow? Maybe she wouldn't care, or might even see your departure as an opportunity? Be careful about threats—if you really don't want to leave, bluffing is dangerous. What would you do if they called your bluff?

The second reason to consider leaving is that your current institution isn't allowing you to develop professionally or personally. It is natural, then, to consider moving someplace that offers better opportunities—stronger programs, better access to graduate students, a more appealing climate (physical or social), and so on. That's fine—it's *your* life. Often the time it is easiest to move is after you've been an assistant professor for three or four years—you can still apply for assistant professor positions, but you will have a more established record that makes you more competitive.

The third, and best, reason for leaving is to be with someone you love. Long-distance relationships are hard—but for the right person, commuting is worth it, as is uprooting to be together. I'll vouch for both![12]

Being an assistant professor is a trial period—for both parties. If, in your first years, you see that this isn't the right relationship, start looking for other jobs. It will only get harder if you wait until after tenure. But avoid burning bridges behind you—reputations travel. You may hate your colleagues, but aim for an amicable divorce.

The best way to stop being an assistant professor, of course, is to get tenure. And despite the anguish and stress that the tenure process is almost certain to evince, most people succeed. Most assistant professorships do not end with tears, but with champagne toasts. I hope that is the case for you.

---

[12] Gwen and I commuted between Fairbanks, Alaska and Berkeley, California for five years before finally moving to Santa Barbara together. That was over twenty-five years ago.

3
Success
Tenure

> *Now this is not the end. It is not even the beginning of the end. But it is, perhaps, the end of the beginning.*
>
> **Winston Churchill**[1]

> *I've got tenure. How depressing.*
>
> **Kathryn Blanchard**[2]

3.1 Tenure: Surviving

Tenure is a strange and often terrifying institution—it's up-or-out and your career is on the line. But tenure isn't a finish line you can stumble across and then collapse. Tenure merely means you've proven yourself (again) as being qualified to be part of this community. If you drop dead like the original marathoner in ancient Athens, you're no use. A tenure case isn't merely about whether you've published enough or brought in enough money—it's about trajectory. The real question underlying a tenure case, as with all personnel cases, is about the *next* step: Are you on a trajectory toward full professor?

The expectations for tenure vary. At some of the most elite private universities, granting tenure can be almost unheard of. (I've heard tenure tales that make Hannibal Lecter sound warm and fuzzy.) In contrast, when I was at the University of Alaska, I felt that as long as I was breathing, I needn't worry—expectations were modest (although many faculty were outstanding) and it was nice not having the stress. Most places though, including the University of California Santa Barbara (UCSB), have high standards, but we hire with the hope, and expectation, that faculty will thrive. Our standards are high, but we invest in recruiting talented people, through time and emotional energy,

---

[1] November 10, 1942, following the British victory at El Alamein.
[2] Blanchard, K.D. (2012).

Your Future on the Faculty. Joshua Schimel, Oxford University Press. © Oxford University Press 2023.
DOI: 10.1093/oso/9780197608821.003.0003

and by strong start-up packages. This seems the pattern at most universities; tenure is the rule, not the exception—most people succeed.

But, as I noted, a tenure vote is really about *trajectory*. If someone isn't on a path likely to lead to promotion to full professor, it's probably better to deny tenure. When someone stalls at associate it likely represents a multiple failure: by the individual, who quit pushing, and by the university, who tenured them but didn't support their further growth.

When a department evaluates a tenure case, they look for evidence that you're solidly on that trajectory. For those in teaching positions, that evidence may include classroom visits by departmental colleagues, student teaching evaluations, assessing curricula and course materials, teaching awards, and so on. In a research university, where scholarship is the primary metric of accomplishment, the key evidence is usually the letters we get from your peers and colleagues around the nation. "The letters carry the case" is largely true. As I've noted, our department mates are often not in our subfields and may lack the firsthand expertise to judge our records. More importantly, perhaps, the campus agencies that review our cases know better than to fully trust what our departments say about us! There is an inherent conflict of interest involved in evaluating our colleagues—we have friends (or enemies), collaborators, and so on. Such relationships get in the way of judging one another objectively. Strong letters that describe your contributions, accomplishments, and potential are critical—we trust your academic peers who know your work.

3.2 Life After Tenure

A senior colleague of mine once said that associate professors are the happiest people in the academic ranks—they are between the stress of a tenure case and the growing administrative and leadership responsibilities that often come with full professor. You're in a window where you can focus on the fun stuff: scholarship, working with students, and teaching classes.

But against that is the lead quote from Kathryn Blanchard, and extensive research, notably by the Harvard-based Collaborative on Academic Careers in Higher Education (COACHE; Matthews 2014). In fact, associate professors aren't the happiest academics—they are generally the *least* satisfied with their careers and institutions. Multiple metrics highlight associates' dissatisfaction with their life balance, the recognition they receive, their departments and universities, and with a suite of other variables as well (Jaschik 2012).

I might hypothesize that associate professors, as a group, are dissatisfied because some of the people who took those surveys had been associates for a *long* time. Am I surprised that someone who's been an associate professor for twenty years might be dissatisfied with their lot in life? Never moving into the higher salaries that come with promotion to full professor? Watching new people come in and advance past them? The COACHE data do support the idea that "stalled" associate professors are particularly unhappy—people who'd been associates for more than six years were significantly less satisfied than those who had been associates for less than six years. But notably, even people who had been associates for less than six years, and so presumably were on track toward promotion, were still less satisfied than either assistant or full professors, falsifying my hypothesis.

Rather, there appear to be several phenomena that lead to lower satisfaction among associate professors than among either assistant or full professors. First, in most careers job satisfaction follows a "U" shape. People are happy in their jobs early on when they are flush with idealism and enthusiasm. But those wear off, while the challenges of balancing work and family grow, dampening overall job satisfaction. But then as we age further, we gain stature in our institutions and financial stability at home; career satisfaction often reflects that and recovers. For academics who have children, the associate, midcareer, stage is also often the period of struggling to find time to read things written by someone other than Dr. Suess, and to write with a keyboard instead of a crayon.

Add in that you just "won" the race and are holding the trophy. You achieved a goal that's been the focus of your life for a decade or more—getting your Ph.D., getting a job, *getting tenure*. But transitions, even good ones, can be emotionally trying, even disorienting. Now what? Who am I? In Chapter 1, I quoted from *Writing Science* about finishing my Ph.D.: "The degree I'd struggled for meant nothing; it was just an entry ticket to this new arena, where I had to start proving myself all over again." Well, guess what—it's the same with tenure! Also, as Blanchard noted in her article, "there are less-savory synonyms for the pleasant-sounding euphemism of 'job security,' such as 'stuck' or 'trapped' or 'you'll never get out of this godforsaken place!'" Such feelings naturally lead to frustration or even depression. Realizing that you might have another 5,000 final exams to grade before you retire could easily push you over the edge!

I might also argue against the idea that associate professors are the happiest because it's before they face the service loads expected of full professors. For many that just isn't true— departments may protect assistant professors by limiting departmental service loads, but that protection evaporates with tenure. And although campus-level activities (e.g., academic senate committees) can be vastly time consuming, they are things we say "yes" to, or often even *volunteer* for. My experience is that those activities aren't *fun* the way playing with a new data set may be, but they are often personally rewarding.

I will hypothesize that being an associate professor is harder in book-culture (i.e., humanities) than in article-culture fields, particularly than in the natural sciences. For a natural scientist, being an associate means doing more of the same—new projects, new science, new papers to write. But, it's at a time when your reputation is growing, you've built a research group, and the "momentum cycle"[3] is spinning. Life can be pretty good, and the hurdles we face (e.g., getting a paper accepted) are more speed bumps than road blocks.[4] In book culture, your first book, which gets you tenure, often grows from your Ph.D. research. Evolving a dissertation into a book is a major task (Germano 2013, 2016), but it's still often research you've been focusing on for years. Now it's time to develop a new project—but without a Ph.D. advisor to help, and while you're teaching several courses each semester, managing service work, and dealing with home life.

This hypothesis stands up better than my earlier one, and not just from anecdotes. At my own university, the academic senate committee on academic personnel has done several studies exploring whether faculty in the book disciplines might be disadvantaged within the University of California personnel system.

I'll offer additional hypotheses about the post-tenure dip in job satisfaction. First, that it might be more intense in undergraduate institutions than in research universities. With generally smaller departments (and so less protection from administrative work), and less support for research, recharging your intellectual batteries might be harder. I'll also hypothesize that it is harder for women, on whom pregnancy and the demands of child-rearing fall most heavily and typically in the midcareer years.

Given all that, it's easy to see how crossing the tenure "finish line" could pull the trigger on a mid-life crisis. The "sausage factory" of academe is no longer a hidden system that just makes tasty sausages. You've learned how messy the inside workings can be, and you've seasoned the sausage with your own blood, sweat, and tears!

---

[3] See Chapter 2.
[4] At least if you overlook that success rates with some funding programs have dropped below 10%.

3.3 OK, So You Got Tenure: Now What?

Academe may be similar to other career paths where a midcareer dip in satisfaction is common. But as long as we make it through that and get promoted, we generally recover. The workload may not get easier, but the sense of satisfaction and increased recognition can make the job more rewarding.

Up until tenure, we are on tight timelines—four years for college, five to six years for a Ph.D.,[5] six years as an assistant professor. Tenure may be the first time we've been able to stop to catch our breath.[6] Whether you feel it as "job security" or as "trapped," your job is now safe, at least as long as you don't get caught in a gross violation of professional ethics, criminal laws, or both.[7] Tenure provides the chance to think about the directions in which you are going. That doesn't mean to *slack off*—you didn't get promoted to do that. But you've proven and established yourself; now, you get to think about where you are and where you're going. Do you want to keep rolling down the track you're on? Alternatively, do you want to shift focus and define yourself in new ways? New research directions? Shift toward leadership positions? It's *your* university and you likely have several decades to play with ahead of you. What are you going to do with them?

3.4 Looking Forward

Here is where the Churchill quote applies. Tenure is not the end; it's not even the beginning of the end—it truly is just the "end of the beginning." Tenure merely demonstrates you've proven a level of professionalism and proficiency, and that you are now ready to take on the hard work ahead. That work doesn't get easier or less painful—research is still research, teaching is still teaching, and universities are still infuriating. But you've met their measure and learned how to cope.

Looking ahead at how to thrive in your career, the first step is to recognize, and accept, those truths—tenure is just the end of the beginning. But you're not an assistant professor anymore. While you were an assistant professor, you were probably protected from some administrative service and also may have had reduced teaching. Your university gave you the opportunities to

---

[5] In the United Kingdom and Europe, a Ph.D. normally takes three to four years, while in the United States it can sometimes take even longer.

[6] Promotion to full professor would normally be another six-year term, but that's not "up or out."

[7] To fire a tenured professor, it often seems to call for "both." For example, sexual harassment is both illegal and a violation of professional ethics.

succeed—or at least to make it hard for you to blame them if you didn't. That ends with tenure.

Universities have historically operated on an implicit assumption that as a tenured professor, you know everything you need to know about being an academic and about how to thrive in your department, your university, and your field of scholarship. Your training is complete and you're ready to fly on your own, without additional support or mentoring. Anything else you might want or need to learn, you can do on your own. It's an interesting idea, but it's rubbish. If it were true, why would there be *three* ranks in the academic hierarchy? Nope, following tenure, there is at least one more promotion in your future—academia really does understand that your development remains incomplete! We just don't provide any structure to help you with that further development.

This is something militaries have long recognized: midcareer professionals still need further developmental training. Army majors go to Command and General Staff School to prepare for the responsibilities they will take on as field grade officers; emerging leaders may later go to War College to develop higher-level strategic skills.[8] In academe, we have nothing comparable. During our Ph.D.s we focus on becoming scholars and getting trained to do so. As assistant professors, we polish those skills and we hope to develop new ones in teaching and mentoring. But we figure out what we're doing either by making it up as we go, or by watching colleagues and emulating (or avoiding) their behaviors. Some departments organize mentorship systems for new faculty,[9] but we typically have little formal coaching and likely none post-tenure.

Universities are recognizing that this lack of midcareer support and training is a mistake (Baker 2019). Offering such coaching appears particularly critical for women and minoritized faculty, who are generally less clear about the trajectory toward promotion, and who frequently get asked to take on a disproportionate level of service roles (Kulp et al. 2019). What activities should we take on? How much? How much is too much? How do we even figure this out?

As an assistant professor, the answers are at least fairly straightforward because your job is to keep your job. In a research university, that means research first and teaching second[10] while at a teaching-focused institution, teaching comes first and foremost. You should do enough departmental and

---

[8] Majors are comparable in age and place in their career trajectory to assistant professors. The Navy's analogous training programs are the *College of Naval Command and Staff* and the *Naval War College*.
[9] For example, some UCSB programs appoint a senior faculty member as mentor to each new hire. Others leave it to new faculty to find mentors.
[10] How far second may vary between universities or even across departments, but for sure *don't* let teaching become a problem that raises a Red Flag on your file.

campus service to check that box and to engage on topics you care about. But "too much" is pretty easy to define—if it measurably interferes with your research or teaching, it's too much.

Following tenure, the answers become fuzzier because there is no single clear metric. The primary criteria for advancement will still be research and teaching, but you get to make choices about your activities, and to pay the price for those choices—but that price won't now include losing your job. What do you want to do?

Some years ago, I was invited to become a Chief Editor for *Soil Biology & Biochemistry* (SBB). I was already on the editorial board, but that meant reviewing a few papers a month. Being a chief editor meant handling review and decisions for ~200 papers a year. I asked my department chair her opinion. She noted I'd already checked the "chief editor box" on my CV by serving in that role for the Synthesis & Emerging Ideas section of *Biogeochemistry* (which only involved a few submissions a month). How would taking on the SBB job help at my next promotion?[11] Beyond, of course, involving a lot of work (ca. a day a week) that would distract, and likely *detract,* from my own research and teaching? I'd be crazy to say yes! Yup, she was right—it was a terrible idea. I ignored her. I decided I wanted to do it—I find editorial work rewarding and a valuable contribution to my community. Now fourteen years later, I'm one of the two coeditors-in-chief, responsible for managing the entire journal! Has the job taken time I could have spent writing proposals or papers? Of course! I spend some time on editorial work almost every day, and part of my Saturday "ritual" is to clear the SBB in-box. I definitely pay a price for having said yes, and so rationally, I probably should have said "no" then, or quit since. But, I didn't.

That's the type of choice that tenure frees you to make—it allows you to take on activities that might seem crazy from a narrow careerist point-of-view, but are valuable to *you* and that you want to do. I doubt I'd be writing this book now had I taken my chair's advice then. And please don't ask how many hours writing this book is taking, or how many proposals or papers I could have written if I weren't. I really don't want to know. I know my first book had value, and I hope this one will, too. Doing each has felt right when I did it, and I'm an *academic*, not just a researcher or a teacher. With tenure, I get to define what being an academic means for me.

Having the freedom to define yourself, however, is a major part of why gaining tenure can be confusing and challenging. You could easily have

---

[11] I was Full Professor, but the UC system has two more full-career review "promotion" steps after that: Step VI and Distinguished Professor.

another thirty years of your career ahead of you—what are you going to do with that time? The question needs an answer. Please don't dream, though, that I'm suggesting that it needs the answer *right now*. That would be absurd—my career has gone in directions I'd never have imagined in my most fevered dreams or wildest fantasies. So, no, you don't need all the answers now, but maybe for the first time in your professional life, you get to pause and think about them, because you no longer have to focus on immediate survival. Do you want to keep doing more of the same? Do you want to try out new or different roles? No one else can answer those questions for you anymore—and their best advice might not work for you.

I'll end with a quote from Anne Lamott's *Bird by Bird*: "It's teatime and the dolls are on the table." What are you going to do with them?

4
Thriving in Academe When You Are Not a Heterosexual White Man

> *There is no victory in winning tenure if you sacrifice your core self in the process.*
>
> K. A. Rockquemore and T. Laszloffy

Once upon a time, almost all university faculty were white and male. That has changed over the last decades. Women now comprise half of assistant professors, almost half of associates, and about one-third of full professors. The pattern is similar for faculty who do not identify as white, who now comprise 29 percent of assistant professors, 23 percent of associates, and 18 percent of full professors (Hussar et al. 2020 [Indicator 2.7]). Some of the drop-off with seniority is just time lag—women and minorities have been a growing fraction of the academic population. Whereas in 1976 women accounted for just over 20 percent of doctoral degrees, they now account for over 50 percent; in parallel, in 1976 fewer than 10 percent of doctorates went to people of color, a number now over 30 percent. Given the life span of an academic, it takes decades for a change that starts with students to percolate up through the full professor ranks. However, some of the drop-off in numbers between junior and senior ranks reflects higher attrition among faculty who are women or from minoritized groups (Institute of Medicine 2007 [Chapter 3]).

Academe and society more broadly need our university faculties to continue to diversify and to reflect better the population of the nation. That requires several things, including increasing opportunities, reducing obstacles, and recruiting a more diverse professoriate, but it also requires that those faculty who are hired succeed.

It would be, however, hopelessly and arrogantly inappropriate for me to offer personal advice on this. The issues I've discussed in the last chapters are things all of us face as academics, and things with which we all must deal. But being an academic is harder if you are not a heterosexual white man. Women may have to deal with sexual harassment as well as pregnancy and childbearing. People of color and people who are LGBTQ + still too regularly suffer isolation and discrimination.

As a university community, we must minimize barriers to academic success—I'd like to say we must eliminate those barriers, but I suspect that may be a Holy Grail beyond the capacity of humankind. Still, we must try. This underlies the idea of antiracism as expressed by Chaudhary and Berhe (2020):

> Avoiding racism or stating that one's lab is "not racist" adopts a neutral stance in a struggle that inherently has no neutrality.

The same can be said for sexism, homophobia, anti-Semitism, and every other form of identity oppression—there is no room for "neutral." As long as these evils exist, they will bedevil their victims and our communities. That is not acceptable.

Systemic problems, unfortunately, persist in the Academy. In the United States, the majority of undergraduate women students report experiencing harassment at some point in their careers (Cantor et al. 2019), and many women and people of color experience hostile workplaces—microaggression, sexual harassment, overt racism—enough that many consider leaving academe—and some do leave. Women, LGBTQ+, and scholars of color remain less likely to get invited to author review articles, serve on editorial boards, and to have the opportunities to take on roles that establish prestige and serve as pathways to leadership (Marin-Spiotta et al. 2020; Jackson 2021). These issues become more severe for academics who live at the intersection of several categories: women of color, for example (Rucks-Ahidiana 2021). Marin-Spiotta and colleagues (2020) note many of the challenges and criticize the "leaky pipeline" model to explain why we have not made greater progress in diversifying academe despite decades of effort.[1] They argue that the metaphor implies that the leaks are *passive*—rather than recognizing that *active* obstructions remain that push women and people of color out of academia.

There are, however, well-documented and proven methods to reduce some systemic obstacles: providing search committees with training in avoiding implicit bias, establishing more defined rubrics for initial evaluation, removing as much identifying material about the individual as possible from

---

[1] For example, the University of California's Presidential Postdoc Program, whose purpose is to "encourage outstanding women and minority Ph.D. recipients to pursue academic careers" was launched in 1984.

applications, or, as they do in auditions for orchestras, having performers play behind a screen so that the judges can only hear the music. Universities and professional organizations must continue to work to make discriminatory behaviors of all types both socially and professionally unacceptable, and punishable if severe enough.

Others with more knowledge and more expertise have discussed these issues, both documenting their magnitude and offering approaches that our institutions can and should apply to reduce the obstructions in recruitment, retention, advancement, and promotion for women, people of color, and other marginalized groups. I refer to some of these sources in this chapter. I encourage all to read them and to do what we can to act upon the advice to improve the climate in our own groups, departments, universities, and professional societies. Ultimately universities change from within as we improve our culture to reduce the obstacles to inclusion, and doing so is something all of us must work on (especially those of us who are white, heterosexual, men). This requires that all of us openly acknowledge the problems that we are part of, but as a changeable baseline, and to guide our efforts to improve.

But this book isn't about all that's wrong with the academy—general kvetching is something to do over drinks with friends, but if someone specific has treated you badly, consider your university's formal charges processes. Criticism is valuable to identify problems, but it is only a first step. The focus of this book isn't even on how we, as institutions, can eliminate the problems; others are more qualified than I am to offer those solutions. Rather, this book is targeted toward *individuals*, and I aim to offer insights into how to succeed within the academy. If you do that, then you can become a leader and work to solve our systemic problems.

As I noted, however, it would be inappropriate for me to try to offer specific thoughts or advice from a personal point of view on how to overcome these challenges to thriving as an academic when you don't look or love like me—"too male, too pale, too stale," to steal words from a lecture by Larry McEnerney.[2]

One of the best resources I know that can offer those perspectives is Kerry Ann Rockquemore and Tracey Laszloffy's book, *The Black Academic's Guide to Winning Tenure—Without Losing Your Soul*. The focus of their book may be Black faculty, but most of their advice should be relevant to academics from other groups as well. Even if, like me, you are "male and pale" it's deeply insightful and offers excellent general career advice. But beyond that, it offers

---

[2] McEnerney is the Director of the writing program at the University of Chicago. The lecture is available at: https://www.youtube.com/watch?v=vtIzMaLkCaM

insights from the perspectives and experiences of people who are not "me," and so may help us all create a healthier, and more functionally diverse, university. Second, and closely related, is the National Center for Faculty Development and Diversity (https://www.facultydiversity.org). The NCFDD (which Dr. Rockquemore founded) provides an array of materials, from weekly "Monday Motivator" e-mails, to videos, webinars, and workshops on how to succeed as academics. NCFDD provides a virtual community and excellent advice; their focus is on junior faculty, but their material is useful for scholars at all levels. If your college or university is not already a member organization, encourage it to join.

Beyond that, I invited several people who aren't "me" to offer their thoughts on how they have overcome specific challenges to become successful in academe. I picked their brains and have encapsulated their words. On some other points, there are resources that already develop advice at length and I note those sources. I've organized that input into several thematic areas. I would like to have engaged more voices and wisdom, but people are busy and this is a delicate subject.

4.1 Learning to Say No (Redux)

Go back and read the section on "saying no" in Chapter 2. Now practice. Repeat the mantra: "No is not spelled with a K." But then see the section in Rockquemore and Lazloffy on "The N-Word":

> With the best of intentions, white faculty often advise faculty of color to just say no. At the surface level, this is exactly what you must do. But at a deeper level, this seemingly simple advice too often fails to acknowledge the sheer volume of requests that the average black faculty member receives over and above majority faculty . . . saying no is more complicated for black faculty than it appears. Take time to understand why you say yes to service requests.

There are good reasons we say "yes" to things and even to volunteer for some. So although it's easy to advise people to "just say no," it's never that simple, and of course you shouldn't say no to *everything*. Rockquemore and Lazloffy note compelling reasons to say yes to some things: trying to make up for the systemic failures you may have faced as a student, fear that you might be punished if you say no, feeling a debt to the person who asked you, and so on. How to respond can involve a complex calculus. Importantly, they advise not to make a snap call—give yourself time to work through the decision and to come up with your response. They also offer a list of ten diplomatic ways to say no.[3] They also note that if service undermines research, writing, and teaching, "you will not be at your institution long enough for your efforts to matter." View your career "as a book with many chapters. If you give yourself permission to focus in your tenure-track years on what is going to win you tenure and promotion, be assured that you are not eliminating other goals but instead are spreading them out over your career."

This parallels my arguments in Chapter 2 that your job as an assistant professor is to keep your job, and in Chapter 4 that tenure is merely the end of the beginning. Tenure means you've earned your place and proven your abilities; after that you should have decades of your career to look forward to. Also remember "The Speed of Academe": you *can* make change in your institution, but supertankers don't turn on a dime, and you need to keep leaning on the tiller to hold the turn and make the change stick.

4.2 Constructive Conflict

Changing university culture and operating approaches will always be delicate and involve conflict. There's nothing wrong with that. Conflict, handled well, can actually strengthen a unit, and equally, can advance your standing within it. But conflict, handled poorly, can be a lose–lose situation, weakening both your unit and the respect your peers may hold for you. Rockquemore and Lazloffy have a chapter on constructive conflicts, in which their critical argument is:

> [T]here are two important, yet often contradicting goals, goals to achieve and that you will have to find a way to be faithful to both. The first is that you **win tenure**, and the second is that you must be **retain your soul** in the process.

They also note that: "[Y]ou will face people and situations that challenge you." And that the "key issue facing black academics . . . is how to negotiate conflict constructively."

We are often stuck with departmental colleagues for decades; we will inevitably have different opinions and priorities. Building a healthy and productive community doesn't mean avoiding conflicts—it means developing approaches, where possible, to avoid letting them become personal and evolve

---

[3] I'd copy it here, but plagiarism lawsuits are ugly, and you should buy their book, too!

into enmity. Once you start down that path, it's hard to come back. And potentially dangerous if that person will vote on your tenure case.

Achieving your long-term strategic goals calls for being thoughtful in your tactics—each small "win" gains you stature and credibility that grants you more influence for later conflicts. Over time, your position within your program will improve, and your ability to take on yet larger battles (and thus make change within your institution) will increase.

Preventing intellectual sparring from mutating into personal anger and enmity has two components—the first is choosing your battles, while the second is fighting them constructively. As a tool for deciding whether to take on a fight, Rockquemore and Lazsloffy recommend evaluating both the short-term and long-term gains and losses for each option. They discuss tactics for fighting the battles you do choose to take on—these include being clear about your goals, making power differentials overt, finding common ground, using "I" statements, and several others.[4] Even if you lose a specific battle over a particular issue, using such tactics to fight "well"—collegially, respectfully, and thoughtfully—establishes your credibility and stature and so positions you to win future battles. Ultimately living in a department and making fundamental change calls for a long-term integrated campaign, not a string of disconnected battles.

I know people who feel the need to throw themselves into every battle, and they would probably justify that on the opening quote for this chapter: "There is no victory in winning tenure if you sacrifice your core self in the process." They'd probably argue that letting a battle pass would be sacrificing their core self. But when you throw your full strength into every battle, you wear that strength away; that is, instead of gaining credibility and stature in your community, you erode it. To win your battles, you need your colleagues to listen when you speak. Remember—as an assistant professor, your job is to keep your job. If you focus too much of your time on political battles, instead of teaching and scholarship, "you will not be at your institution long enough for your efforts to matter."

That is not, in any way, a call to be quiet and patient—rather, it is advice to be strategic in how you approach your battles. As an analogy, I'll reach back to my youth in the 1960s, a time when both the Vietnam war and the Civil Rights movement were raging. The U.S. military fought every battle in Vietnam, but ultimately lost the war, leaving American society badly damaged. During the

---

[4] Remember what I said earlier about plagiarism lawsuits?

same period, Martin Luther King led in the campaign for civil rights. He was brilliant in his approaches, being selective as to when to be combative and when to back off (for now), and in building alliances and support. As a result of his steadfast goals, but flexible approaches of nonviolence, resistance, and protest, he not only achieved great gains, but earned enormous respect that empowered him for future battles. Those fights were far from complete when he was murdered, and we still have not reached the "promised land" he saw before his death. But although we all need to win *the* war, you, individually, still need to win *your* "war"—you need to succeed and thrive in your career. We only get to keep fighting if we survive the battles we fight. That calls for choosing your battles and being effective and constructive in those you take on.

4.3 Find and Build Your Personal Support Networks

Part of both surviving academia and winning your battles is building your support networks (Carver 2017). We all need professional friends and colleagues with whom we can let our hair down, open up, commiserate our struggles, and celebrate our successes. We need to be able to get advice from people we trust and with whom we feel kinship. An excellent book that discusses one model of how such a support network can be formalized is Ellen Daniell's *Every Other Thursday*. She describes the development and dynamics of *Group*, a group of women academics in the San Francisco Bay Area, who first came together in the 1970s for mutual support in their stressful academic careers. Over the years they gathered, unsurprisingly, every other Thursday, to work through their individual and shared challenges. She discusses the strategies they developed to make *Group* work and continue to work over many years, even as individual members came and went. One important aspect of *Group* is that the women weren't all in the same academic department or even at the same university—the challenges they faced were common across all of academia.

A colleague noted to me that she knew of several institutions that have tried to help their faculty develop *Group*-type groups (not just women, but also minoritized scholars), but of course, ultimately, such activities work only when participants feel the need and the value for the group and take ownership of it (Glessmer et al. 2015). Communities grow from the ground up. But just as a gardener fertilizes and provides trellises for their plants to grow up, universities can and should support and facilitate the growth of support communities.

4.4 Dealing with Lack of Respect

One of the ways in which various "isms" can influence academics is in how faculty who are women or individuals from minoritized groups (or both) may not be taken seriously, or have to work harder for respect—to overcome a presumption of whiteness and maleness. For example, when many of us see just an author's last name, we default to an assumption that the person is a man. In a personal example, I learned the Michaelis-Menten equation in high school. It's the basis of enzyme kinetics, describing the relationship between substrate concentration and the rate of an enzyme-catalyzed chemical reaction; it's fundamental in biochemistry and biology. But when did I learn that Menten was *Maud* Menten?[5] Maybe ten years ago? I'd never thought about it before that, but I'm sure I would have assumed that Menten was a man. The presumption that a "scientist," a "professor," or a "doctor" is a man and white remains largely pervasive (Finson 2002).

That "presumption of respect" plays out in other ways. For example, my standard uniform for conferences is black Levi's, a button-down shirt, and New Balance walking shoes that don't look quite like sneakers. I know I can look casual, or even scruffy, and still be taken seriously—I've attended conferences wearing jeans and a T-shirt. Most of my colleagues who are women or people of color don't feel they can assume the same, particularly when they are starting out. They feel a greater need to look "professional" if they expect to be taken seriously.

We can change rules and systems, and we have some tools to change people's behavior, but there are no simple solutions to changing what is in someone else's head. Changing attitudes and assumptions happens subtly, and over time, but can be accelerated by a variety of methods. One is simply to not let people get away with negative presumptions about you. One of the best examples of that I've ever heard was described in a Tweet by Dr. Jessica McCarty:

> At a NASA Earth meeting 10 years ago, a white male post doc interrupted me to tell me that I didn't understand human drivers of fire, that I def needed to read McCarty et al. Looked him in the eye, pulled my long hair back so he could read my name tag. "I'm McCarty et al."[6]

---

[5] She was an extraordinary woman who earned her M.D. in 1911 and her Ph.D. in 1916!
[6] https://twitter.com/jmccarty_geo/status/1361332337678639107

I gather that in response to that story, many other women came forward with similar experiences. Faculty of color have parallel experiences. I might hope that that postdoc wouldn't make the same mistake a second time. Calling out bad behavior, particularly in a way that is unlikely to invoke a confrontational reaction, is a powerful way to make such behavior unacceptable, and to drive it underground, where the underlying attitude may hopefully shrivel up and, if not die, at least fail to spread. Over time, that can create space for new attitudes to become dominant. We have seen some successes—universities don't look or behave like they did in the 1960s! But we still have a long way to go—there are still battles to fight and campaigns to win.

4.5 Master Maternity Leave

Many of the issues I've drawn from conversations and readings are relevant to many groups of people. How the dynamics play out might be different, or feel different, among minoritized communities in academe—women, African-Americans, Latinx, recent immigrants, and others—but they share parallels in how they operate. One issue, however, that works differently is *childbearing*. Men may be *parents* and struggle with the challenges of balancing career and raising children, but men don't have babies—we don't go through pregnancy and childbirth.

Once upon a time, women academics who wanted to have children got a clear message that they should wait until they'd built their career to the point that their tenure case was secure. Over the years, gratefully, it has become institutionally acceptable for women to have children earlier—as assistant professors, postdocs, or even as graduate students—when it suits a woman's wish to start a family.[7] However, although it has become socially acceptable to run an academic career and a family in parallel rather than in series, it hasn't made it humanly easier to juggle. Hence the advice from Dr. Elisabeth Holland: *master maternity leave*. Find out all of your institution's policies related to maternity leave—how much, paid versus unpaid, when can it start, are there childcare facilities available to you through the institution, and so on. What are the policies that apply to your partner—for example, do they get parental leave? Determine this early and get your name on waitlists for access to university daycare—it would be surprising if your university has enough to meet the need (assuming they have any). Policies vary across institutions

---

[7] And incidentally, when women's bodies are biologically more suited to having children than waiting to have a first child in your forties.

and also for different positions within the institution. As I note in Chapter 6, postdocs and other temporary employees are likely to find both the least support and the least clarity—so talk to your advisor about your plans—official policies might not cover everything.

4.6 Being an Active and Positive Member of the Academy

Academe needs to diversify, and that means we need faculty who are not white, male, and so on to succeed. We need you to reach your potential, make tenure, get promoted to full professor, and grow into leadership positions. Most of us want to support you in those endeavors. But as I noted when I discussed "saying no" in Chapter 2, some of our best efforts to win diversity battles in the short-term (such as inviting you to serve on campus committees) might be counterproductive in the long-term—winning battles instead of wars. How to balance such competing institutional priorities will always remain tricky. After all, you have to fight some battles to win the war—but which ones and what is the best way to fight them?

These are battles, however, that our minoritized colleagues should not have to fight on their own—in fact, they shouldn't have to fight them at all! Those of us who are "male and pale" must be on the front lines, which poses the question to us: How can we take stronger leadership roles and be more effective allies in doing the work to create a more diverse and equitable university? Our colleagues still report incidents of bias, microaggression, harassment, isolation, and discrimination. They struggle with always having to be the ones to call it out. Being a good ally means supporting and embracing campus and departmental initiatives to improve our culture, but it also means taking responsibility and action when we see someone treating a colleague poorly (Berhe et al. 2020). That could be as simple as noting, "Uh, Joan just suggested that two minutes ago" in a faculty meeting. At the other extreme, it could mean directly intervening to stop an act of overt racism that we observe, or even filing formal charges ourselves. But minoritized faculty also routinely report being gaslit by well-meaning colleagues and administrators who fail to address the negative effects of exclusionary behaviors. Worse, when they do report exclusionary behaviors, sometimes the tables are turned back on them in a defensive tactic referred as DARVO (Deny, Attack, and Reverse Victim and Offender [Harsey and Freyd 2020]). Such behaviors not only inhibit victims from coming forward to report harmful, or even illegal, behaviors, they also damage the victims. We must do what we can to prevent exclusionary behaviors and to hold accountable those of us who are guilty of them.

There is much more that could be said, but I leave it to others better qualified to say it. I felt it critical to include this chapter, knowing it was nearly impossible for me to write adequately. I'm sure I haven't done the issues full justice, but not saying anything would be the greater injustice.

5

Non-Tenure-Track Teaching Faculty

It's not for the faint of heart

Dr. Helene Gardner

Most university faculty are not professors in research universities, institutions whose mission integrates undergraduate teaching with research and graduate training. Most U.S. faculty are either professors in colleges and undergraduate universities, or are non-tenure-track teaching faculty. In California, although the University of California gets the glory, with its campuses consistently ranked among the world's best, the California State University System teaches nearly twice as many undergraduates (roughly 430 K vs. 230 K).[1] In the United States, roughly three-quarters of all faculty are not in tenure-track positions at all (Flaherty 2018). This pattern is common across all U.S. States and other nations as well—many faculty are involved in teaching undergraduates rather than research or graduate training.[2]

Universities have classes that must be taught, so we recruit different types of faculty to teach them, in addition to tenure-track professors. Some of these are one-time appointments, such as when a professor goes on sabbatical and a graduate student or postdoc covers a class to get teaching experience and extend their funding. Some people regularly teach just one or two classes a year—for example, working professionals in specialized areas such as law, entrepreneurship, or journalism. They do it because they enjoy teaching and may use classes to recruit interns, rather than because the money makes a difference to them.

Then there are people who teach multiple classes a year; for many, teaching may be their full-time job. Of these, only about 20 percent have multiyear contracts; the rest are hired year-by-year or class-by-class. In the University of

---

[1] California's master plan for higher education blocks them from offering independent Ph.D. programs.

[2] In other nations, a smaller proportion of people attend college. In the United Kingdom, the proportion of college students in their overall population is 60 percent that of the United States, and less than one-third of their total "academic staff" is classified as "teaching only." https://www.hesa.ac.uk/news/23-01-2020/sb256-higher-education-staff-statistics.

California system, such lecturers come up for evaluation after eighteen quarters of teaching; if successful, they become a "Continuing Lecturer" with an open-ended contract. That review may occur at a time parallel to a professor's six-year tenure review, but the parallels are weak—some job security, but not comparable to tenure.

The University of California system also has a separate category of "tenure-track" teaching faculty, who have the title Lecturer with Potential Security of Employment (LPSOE); after a six-year "tenure" evaluation they get to drop "potential" and become LSOEs.[3] These faculty focus on undergraduate teaching; they have no requirement for research or graduate training. LSOEs are analogous to professors at undergraduate colleges; in fact, there have been discussions about changing the title to Teaching Professor. But that raised an uncomfortable question: If they are "teaching professors," does that mean that other professors don't teach? Don't give a politician that idea! Instead, the Solomonic compromise was to keep LSOE as the legal title, but to *officially* authorize *unofficial* use of the title "Teaching Professor."

As with the last chapter, however, I can't write this one from personal experience because I *am* a tenured professor in a research university. So, I got input from a colleague, Dr. Helene Gardner; she has taught at several institutions in a variety of positions that span the full range of teaching positions. At the University of California Santa Barbara (UCSB), she started out as an adjunct lecturer teaching one class; now she's "tenured" as an LSOE in the Environmental Studies Program.

I met Helene when I was chair of Environmental Studies. We needed an instructor for our environmental toxicology class; ideally someone with experience in both real-world applications and college teaching. That described Helene, who was looking for new opportunities that would be a step up from what she was doing at the time—including unhappily teaching online college. I felt I was offering a win–win that might lead to greater opportunities; she accepted the offer. So I passed the hiring paperwork to our department staff. Her description of the interaction agrees in fact, but differs in flavor:

> I was asked by the chairman of the Environmental Studies Program (ESP) if I would be willing to teach their course in toxicology the upcoming quarter. I was. I had a very good idea of what to expect as an adjunct and was not disappointed. I signed a contract to provide instruction for ten weeks and submit grades by the grade deadline and, in turn, was paid a small sum of money. There were no benefits, I had no

---

[3] L(P)SOEs and LSOEs are members of the academic senate and have salary scales that parallel those of professors.

responsibilities beyond those for my course, and I received no orientation or introduction to the institution. It was as if the university and I were casually dating, so we were free to date other people and there was certainly no need to meet the family.

I think the different tone captures the challenges facing adjuncts and other temporary faculty. Administrators convince ourselves we're providing opportunities (within our limited resources); the people we hire, however, know the realities that come with those opportunities. There is truth to each perspective, so have some sympathy for the department chair who is apologizing for the salary they can offer—they didn't come up with that number. However, you still can't pay your rent with their apology.

Importantly as well, an unstable teaching force produces unstable education. Students run into problems when they need to change a grade or ask for a letter of recommendation but their instructor is gone! To students, every teacher is "Professor So-and-So." They rarely understand, or care, whether the person teaching them is a graduate student, an adjunct lecturer, or a tenured professor. What matters to them is that you are their teacher—you're older and you're giving them their grade. When colleges and universities rely heavily on adjuncts, students suffer from not having faculty to help them and mentor them outside of class. Overreliance on adjunct faculty, therefore, can reduce student success and graduation rates.

In her comments, Dr. Gardner highlighted some positive aspects of adjunct life, but others are distinct negatives. The positives revolved around the flexibility: to teach topics she cares about, to organize her work schedule around her family and outside consulting, and the freedom to teach courses of her own vision and "with eager students of all descriptions."[4] The negatives included poor pay, poor benefits, and poor institutional support. I distilled out some of her most critical points—which she encapsulates as "It's not for the faint of heart."

Her comments largely reflect points I've read elsewhere and heard from other colleagues in adjunct teaching positions. Several themes consistently emerge, which offer guidance for faculty who are working as adjuncts, as well as for those of us who are tenure-track faculty, or who hold administrative or leadership roles at our universities. Recognizing the issues might help improve both the lot of our adjunct colleagues as well as our students.

---

[4] Many of those students, though, are eager because of Helene, rather than because they started the course that way.

5.1 Lack of Respect

Aside from poor pay and financial instability, a lack of respect is probably the most common complaint from teaching faculty. Dr. Gardner may now be a tenured teaching professor, but she still reports that, "I have been told to my face that, as an LSOE, I am a second-class citizen in the academy. I have been treated that way by some of my colleagues."[5]

The attitude she experienced grows from several places. One may well be simple arrogance, and arrogance is always, well, arrogant. It may be common, but it's never OK to be a jerk.

That arrogance, however, is partly rooted in (though not excused by) two things that are real. One is the importance of research and graduate training in maintaining our academic systems. Institutions such as liberal arts colleges and non-Ph.D.-granting universities[6] may do a great job teaching undergraduates, and even masters students, but they still need research universities to train the Ph.D.'s they hire as professors. The centrality of research and Ph.D. training in the academic enterprise leads some to see these as being a "higher calling" than undergraduate teaching, as noted by Dr. Gardner:

> It was George Bernard Shaw (1903) who wrote, "He who can, does; he who cannot, teaches." I get the impression that some of my ladder faculty colleagues believe that and see us as failed ladder faculty. I am sure there are LPSOEs who wanted careers as ladder faculty but were unsuccessful and landed in teaching, fulfilling that adage, that belief, and their view. But that is not the case for all—perhaps even most—of us; some of us have chosen to teach. Perhaps our career choice is so different from that of our ladder colleagues—and it is so human not to appreciate those who are different from ourselves—that they cannot understand why anyone would make a choice different from theirs.

The misperception that teaching faculty only chose that path because they failed as researchers is both toxic and absurd. Helene chose teaching because it's her passion and she's *awesome*. It would be pointless and inappropriate to even ask whether she has the research creativity to have succeeded as a high-level researcher—she *chose* to focus on undergraduate teaching.[7] No one ever questions why I chose to pursue a research career, or whether I took up research because I wasn't good enough at teaching.[8] I'm a good teacher, but

---

[5] If an LSOE is a second-class citizen, would that suggest an adjunct is a third-class citizen?
[6] Such as Middlebury College, where I got my B.A., and the California State University System.
[7] Additionally when she was starting out, academic environmental toxicology was not accommodating for women.
[8] People may ask how I got into soil ecology, but they don't question the validity or value of that choice.

I know I couldn't have succeeded the way Helene has. We pursue our paths because of our passions and personal choices. But that won't stop people from making wrong assumptions about why others do what they do.

The second factor that leads some professors to look down on teaching faculty is that research careers are more competitive than teaching careers. Research isn't harder than teaching, but there are more students than grant dollars. With ~20 million college students in the United States, we need a lot of faculty to teach them (~1.5 million in positions from lecturer to professor [National Center for Education Statistics 2020]). Although we aim for excellence, you won't get fired for teaching that is merely "Very Good," although such reviews won't earn you an outstanding teacher award. In contrast, if you consistently write proposals to the National Science Foundation that score *Very Good*, which a review panel might rank *Medium Priority*—strong but not quite top-tier—you may never get a dime in funding and would fail as a researcher. Between the centrality of research in academe, and the challenge of securing a research position, some professors allow themselves to feel more important than their colleagues whose focus is undergraduate teaching. They're wrong, but perhaps understandably so.

The university mission is *scholarship*—a mission that integrates teaching and research across all levels of the academy. They form a continuum, not a hierarchy. Without well-trained and passionate undergraduates, there would be no graduate students! Some of us focus on one piece of the conjoint mission, others on a different piece, while still others do some of both. Whatever our individual balance, however, together we are teammates. Successful researchers are entitled to be proud of their accomplishments, successful teachers are equally entitled to be proud of theirs. Being successful in your chosen path, however, is never an excuse for being snotty toward those who chose different paths. But regrettably, it's all too human.

A critical factor that fuels the challenges for adjunct teaching faculty is, of course, money. Teaching faculty are rarely paid well. Universities are chronically cash strapped, something with which I'm sure the ragged-robed dons back in medieval Oxford would have identified. There are more classes to teach than universities can afford if they were just to hire professors, particularly in STEM fields where setting up a lab can cost $1 million or more. So universities hire lecturers and adjuncts and offer the lowest salaries they can get away with. We also train more Ph.D.'s than there are professorial positions for, and so create an oversupply; in some fields there is as few as one tenure-track faculty position for every six Ph.D.'s granted (Kolata 2016). Although many of those Ph.D.'s may target industry, some substantial proportion would love to work in a college or university. But we know what economics says about price when supply exceeds demand. So universities hire people to teach on a part-time or temporary basis with salaries that range from modest to scandalous and often lack decent (or any) benefit packages.[9] This explains why lecturers and adjunct faculty have increasingly unionized (as they have at the University of California)—universities struggle to balance their budgets, and even when we know we're paying lecturers and adjuncts miserably, students are still paying the tuition, and so somehow we have to offer the classes. That often translates into treating poorly people who had just recently been our treasured graduate students! We need to be forced to do what we find it hard to do by choice—hence lecturer unions.

Some of the feeling of being a second class citizen arises from adjuncts not being part of faculty governance. Professors may often bitch about faculty meetings, but we have a voice and a vote. For adjuncts, not attending faculty meetings—being recognized as a part of the faculty—makes them less visible and less clearly a valued part of the academic community. I don't know all my professorial colleagues well, but I at least know their faces and voices. Lecturers? Not so much.

Perhaps the aspect of governance some adjuncts miss is having a voice in the structure of the curriculum and in which classes get taught, and how. Given that programs often recruit lecturers to teach introductory classes, the focus of those classes would generally already be well-defined. Tenure-track faculty get to create a class or two in their specialty. The overall topics of those classes may be relatively well-defined to cover critical areas (i.e., plant physiology or medieval European history), but professors have flexibility in framing their classes and how they approach teaching the material. Lecturers may have some flexibility in how they teach a particular class, but it's harder to invest in creating a pedagogically novel and involved class when you might only teach it once.

If you want more visibility and engagement, attend department events and engage in the broader departmental life (seminars, departmental parties and celebrations, etc.), if you have the time. Not all departments, however, are good about organizing such activities. If yours is not, perhaps suggest to the department's chair or business officer that they do more.

For all these reasons, permanent and adjunct faculty are separate categories of employee. They each contribute in vital ways to the overall university mission. But all institutions prioritize their commitment to the permanent

---

[9] You can probably add sexism to that equation: teachers had traditionally been women, and salaries for teaching faculty reflected that. Professors, historically, had generally been male and commanded higher salaries.

faculty—the "sunk investment" to maintain the institutional core. If you can't accept that, you will never be content in an adjunct teaching role.

My university has been hiring more permanent teaching faculty (LSOE/teaching professors), particularly in the natural sciences. That growth is driven by a mix of economics, pragmatism, and ethics. Economics, in that it's cheaper to hire full-time teaching faculty to cover foundational undergraduate courses than to recruit professors, who teach fewer classes and may need research labs. Pragmatism, in that large introductory sequences involve administrative overhead and coordination—it helps to have faculty for whom this is their focus. Also, teaching basic material benefits from pedagogical expertise that many professors lack. Finally, ethics, in that these teaching positions are "mission-essential" and should be treated parallel to professors who lead the research and advanced training parts of the university. For the bulk of lecturer-taught elective classes, however, pragmatism wins out and most pure teaching faculty, at both UCSB and other universities, remain adjuncts or lecturers.

There are many reasons people choose an adjunct or pure teaching faculty pathway—to live in a particular place, to be with their partner, to not have the stress levels that go with being a professor, or perhaps because they love teaching more than research. Other people, of course, find their way into the path because of the quirks that fate dealt them. Regardless, if this is your path, accept the realities, but work to change the unappealing ones. We must try harder to institutionalize recognition, respect, and support for teaching faculty, but that change requires recognizing both the realities of the academy and the value we all bring to it. Enhancing our culture and developing that mutual respect will take work from *all* sides.

5.2 The Confusing Deep End You're Going to Be Tossed Into

Every university has systems we rely on to do just about *everything*.[10] But long-term employees tend to suffer from the "Curse of Knowledge." We know what we know—it's easy to forget that new people don't. "UCSB" has become my native language, but it wasn't when I arrived. How do I issue add-codes to allow students to join a class late (*e-Grades*), manage assignments and post information (*GauchoSpace*), or assign grades (*e-Grades again*)? What the hell is a *Bio-Bib* (specifically formatted CV for merit reviews)? The first time

---

[10] Chapter 6.

you deal with these things, or the first three times, they are opaque and not at all obvious. When they all roll over you, new, at the same time, while you're trying to get a class going, it's overwhelming.

Don't expect people around you to appreciate how overwhelming it is—we already speak the language you're just learning. The learning curve may be steep, but the sympathy curve is shallow. If you ask an experienced professor, they may seem unsympathetic because they forgot just how overwhelming it is—which might increase your sense of not being respected. There are, however, people with the answers, notably the departmental staff who will be your lifeline.[11]

Every department has a similar array of things to deal with. But each department evolves its own culture, habits, and ways of doing those things. Some policies and practices may be coherent across the university, but others will differ among units. Experience in one program might actually be misleading in another. If some people are unfamiliar with the rules that apply to your specific job title, they might also give you misinformation—well-intentioned, perhaps—but misinformation nonetheless. For example, Dr. Gardner notes:

> Every time I have come up for evaluation, I have been told by everyone who gave me advice that I would be evaluated in four areas. It was only when I came up for tenure myself and actually read the Red Binder, I found that I was to be evaluated in only three areas.[12]

When units have different cultures and policies in how they treat adjunct faculty, it will inevitably create inequity and other sorts of problems:

> You might imagine that full-time lecturers in different departments talk to each other, that the inequities among the lecturers in different units are discovered, and that this is a problem. We do, they are, and it is.

5.3 Assessment: How Are You Evaluated?

Closely related to issues of both respect and confusion is that although you will be evaluated on your teaching, we don't have good ways to evaluate teaching. Most universities use student surveys and written comments to

---

[11] Chapter 7.
[12] Professors are evaluated on Research & Creative Activities, Teaching & Mentoring, Professional Activities, and University & Community Service; teaching faculty are not evaluated in research.

assess teaching—but the scholarship questioning the reliability of student surveys is deep and extensive. Hence, don't count on only student comments. Dr. Gardner's advice is to ask a faculty colleague to sit in on at least one of your classes. Get them to write an official letter for your file if they can.[13] She also offers this:

> If they cannot write to your file—and this is important—at least there will be people sitting at the table when your case is considered who know and appreciate the good job you do and can speak personally for what you do in the classroom.

5.4 Negotiate Where You Can: Limited Power Isn't No Power

Even when you're offered the opportunity to teach just a single class, there may be a surprising number of details that you should be aware of and be prepared to discuss, possibly to negotiate. Salary is the most obvious, but also the least likely to be negotiable. Other issues may be more flexible; for example, is there a teaching assistant? If not, would the department pay to hire a student reader to help with grading? Even just several hours help a week can make a huge difference in your life, and given what a reader makes, they're bargains. Also up for possible discussion would be how many students are in the class or when it meets. However, Dr. Gardner notes:

> Your courses will probably have been scheduled before you arrive, so, initially, you will have no control over the size of your courses or when they are offered.

That's likely true, *initially*. The class you're taking over might have been in the same slot for decades. The law of inertia applies: Things will continue as they are unless an adequate force is applied to cause them to change. Add to inertia, the "Law of Laziness." For most faculty, our most limited personal resource is *time*, and this is likely even more true for the program chair. If my department has a class that needs an instructor, I want to get it settled. If there's money in some pot I can throw at it to get you to say "yes" and wrap this up? Do it—get this off my desk!

Once you've done a class once, though, inertia works for you and you gain some leverage. If the chair has someone who does a good job, would they

---

[13] In UCSB's Environmental Studies Program, this is policy—we assign a professor to attend at least one class session taught by each adjunct lecturer to ensure we have external evaluation of their teaching.

really want to find someone new? Someone who might not be as good? Of course not. Dr. Gardner comments:

> It's great to be a team player, but once you're established, you can change what gets offered when to what works best for you.

If you ask to make modest changes to the class, granting your request is probably less work, and less risk, than letting you go and finding a replacement. Inertia and laziness both lean toward keeping you on board. Making some changes in the class isn't such a big deal, unless you're pushing for something impossible, such as a prime-time slot (e.g., 10 AM MWF) when every classroom on campus is taken.

After you've done a class once you not only build personal stature, but you'll also have a better sense of the class. How many hours does it really take? What would it take to make the class that one extra step better? Being concrete is a cornerstone of compelling arguments; you need to give the program chair the ammunition to back your request.

Ask about such issues early in the planning cycle. If you ask, the answer might be "yes." If you don't ask, the chair might not think to offer, but that doesn't mean the answer *couldn't* be "yes." When I recruited Helene to teach environmental toxicology, I didn't know all the logistical details—it wouldn't have occurred to me to know. Some things may truly be non-negotiable or just not doable. But a gentle question should give you a sense of whether there is wiggle room. So ask, and be prepared to offer good arguments for why the answer should be "yes."

As Helene noted, life as adjunct teaching faculty is not for the faint of heart. As important as the job is to the mission, you will almost certainly have to put up with feeling underappreciated, overwhelmed, and powerless, at least occasionally. That might not be any different if you worked for Google, Microsoft, Pfizer, or Wells Fargo, but they'd almost certainly pay more. But if you love teaching and working with students, being university teaching faculty can be satisfying and rewarding. But a last word from Dr. Gardner:

> Being an adjunct works if you remember why it works for you and it is something that you choose to do every term.

SECTION 2
UNIVERSITY SYSTEMS

Amateurs talk about tactics—Professionals study logistics

As faculty, our day-to-day lives are structured by our institutions—the administrative systems and the staff who run operations. Without them, there could be no university. Dealing with our systems may be frustrating at times, but they define the landscape in which we live. Thriving as a faculty member requires an adequate understanding of our systems structure and logic. Equally important is figuring out how to work effectively with the staff members who make our universities function. We rely on our staff, but they are often underappreciated, even though they are our teammates.

6
University Administrative Systems

> *Every quote about administration or bureaucracy I could find is about how they are inherently inefficient or even evil, led by incompetent bozos. So pervasive is this idea that the very OED definition of "bureaucracy" includes the statement "frequently with depreciative connotations."*
>
> *After its technical review by the identified systemwide stakeholders, it was determined that P-196-25 is a procedure and not a policy.*
> **University of California Santa Barbara, Policy Coordinator**

Everybody hates bureaucracy. We may admire, even revere, *leaders*—people who run nations or create successful companies—but we despise the structures they use to run them. Steve Jobs started out building computers in his garage, but you don't manage 130,000 employees out of a garage! As organizations grow, we create rules, policies, practices, and culture. We compartmentalize functions (sales, financial, legal) and we create *bureaucracy* because systems and structures—even policies and procedures—are essential to organize and manage human activity.

We may hate bureaucracies, and disparage bureaucrats, but every month money appears in my bank account, transferred from the University of California (UC). I like getting paid! That money doesn't appear by magic—there is an elaborate system to manage the flow of money into and through the university, and to *me*. But most of the time I can happily ignore it.

It's only *bureaucracy* when it becomes visible; that is, when a paycheck doesn't appear in my bank account, or when we experience "an excessive concern with formal processes and a tendency for administrative power to increase and become more centralized, and hence by inefficiency and impersonality; officialism, red tape."[1]

Equally organic to that definition of "bureaucracy," however, is its first description: "administration by a hierarchy of professional administrators

---

[1] From the *Oxford English Dictionary* definition of "Bureaucracy."

following clearly defined procedures in a routine and organized manner." We're happy when those professional administrators keep the *defined procedures* working in the background to ensure our paychecks appear. But just as with my body's systems that keep me breathing and my heart beating without my having to think about it, we only notice the systems when they fail. Perfection is invisible.

Universities, of course, aren't perfect. They can, in fact, be quite infuriating places to work—our administrative systems are often cumbersome and outdated. Universities are the most complex institutions in civilian society, having more distinct, but interwoven, missions than any other.[2] Companies are about making and selling stuff. Nonprofits are about saving endangered species, social policy, preaching religion, or taking care of people. Hospitals may be mind-bogglingly complex, but they are all about keeping us alive and making us healthy. Universities are worse. Much. The University of California even manages not just a hospital, but an entire hospital *system*, with five separate medical centers. UC also manages national labs, oceanographic research ships, military officer training (ROTC), farm advisors, and a dizzying array of other elements, of which undergraduate teaching is only one. To add to the structural and mission complexity, universities have deep cultural history and traditions going back to the Middle Ages; we live with the accumulated baggage of our histories and development. Why else, in southern California in the Anthropocene, do students still graduate in robes designed for northern Europe during the Little Ice Age?

The other commonly noted aspect of academic bureaucracies is their speed, or lack thereof. The "speed of academe" is legendary. At UC Santa Barbara (UCSB), when faculty go up for merit reviews, we submit our paperwork (CV, publications, teaching evaluations, self-assessments, etc.) at the beginning of the academic year, but we don't hear final decisions until the next summer. Why on Earth does it take a year to decide how big a raise to give a professor?

Dictatorship can operate quickly; we could let a department chair just say, "I like her, so I'll give her a big raise, but he didn't support my pet project last year, so he doesn't get anything." That would be quick. But fairness, transparency, and consultation take time. At UC, a faculty member's department first evaluates and votes on a merit review. Then it goes to the dean's office, where staff check the materials before the dean assesses it and offers their perspective

---

[2] The military is the only institution that competes. After all, they even run their own universities—the service academies (e.g., West Point) grant undergraduate degrees, while both the Navy and Air Force have postgraduate schools that grant masters and doctoral degrees.

on the case and the department's recommendation. Then it goes to the academic senate's Committee on Academic Personnel, which evaluates all these materials and makes a recommendation to the Vice Chancellor for Academic Personnel, who makes the final call. Everyone in this chain is busy with research, teaching, and administrative service. But the system is careful, analytical, and reasonably equitable. Justice, swift but intemperate, may be fine when someone else's neck is in the noose. But when it's ours? Then, we value thoughtful and careful jurisprudence. That takes process, and accepting that systems take time.[3]

The size, scope, and sometimes bizarre culture of our institutions guarantee that we will run into situations where systems don't easily allow us to do what we want, or make it harder to do things than it seems should be necessary. Sometimes, there may be no clear policy on something that needs one! Many of those challenges can be overcome. That might require figuring out how to deal with one difficult person. It might mean figuring out a clever kludge to work around the problem. But it might require creating a whole new initiative or program. Sometimes that's easy, but other times it will prove harder and take longer than you would have dreamt possible. Challenges grow from various sources, and each calls for different solutions.

When making change involves time and energy, we tend to stay with what we have until either it becomes glaringly obvious it's broken, or until some wise person sees that it has passed its "use before" date and is heading toward becoming broken. Can you identify which policies and systems are still fruitful; which are legacy, but harmless; and which are breaking or broken and really should be changed—without the pain of a crisis?

I categorize the sources of challenge into six broad areas, but these areas overlap and interact with one another:

- Policy
- Practice
- Proprietary Systems
- People
- Politics
- Pecuniary Realities, or if I don't force myself to use a P-word—*Money*

---

[3] Keep in mind the parallel to peer review: when it's our paper, we want fast turnaround; when we're the reviewer, we want more time. Sauce for the goose, sauce for the gander.

6.1 Policy

Sometimes constraints result from actual institutional policy. It's the rules. You won't get around that quickly. You have to either change the rules, or you have to get a waiver or an exception to them. Institutions generally don't like making exceptions to policy because if we make an exception for you, how do we justify not making the same exception for everyone? By granting one exception, have we functionally just voided the policy—but without careful analysis of why it exists and whether it's a good idea?

One of my mottoes is "never let the rules get in the way of doing the right thing," but often, following the rules *is* the right thing. We create rules and policies to achieve valuable goals; I can usually see the value in the goal. But when instead, a rule interferes with achieving the goal, find a work-around or get the rule changed. Work-arounds are probably easier.

Changing policies, even when the change is simple or the policy wrongheaded, can take a shocking amount of work, likely over several years. Are you willing to "fight City Hall"? Universities don't change rules without careful thought—and we are, of course, masters of "careful thought."

If you really want to put on your Don Quixote costume and start tilting at windmills, setting out to change a policy, the first thing to sort out is who made the policy. Which entity within the university established the rule or oversees it, and so has the power to change it?

For example, at UCSB, a student's Ph.D. committee requires only three members, but when I arrived, to have one of the three members come from another academic department, you needed approval from Graduate Division. Because I do soil science in an ecology and evolution department, every one of my students needed either Trish Holden (Bren School) or Oliver Chadwick (Geography) on their committees. I had to go through the approval paperwork for *every* student. That wasn't really a big deal—ten minutes to write a brief explanation on the petition form. But it bugged me: UCSB prides itself on interdisciplinarity, so why create an obstacle to interdisciplinary committees? Why not allow departments to decide whether they would accept a member of another department as a member of a student's committee? It seemed a silly policy, and I became allergic to it.

So, I set out to eliminate the allergen: I volunteered to serve on Graduate Council.[4] When I proposed the policy change, there was no opposition, but it still took a year to get input from all and sundry, move the proposal forward,

---

[4] The committee of our academic senate that oversees all things related to graduate programs and training.

and get final approval from the Academic Legislature. I succeeded, but, in the simple balance, comparing the time and energy it took to get the policy changed, relative to that involved in getting student committees approved, it was a mind-blowingly bad investment on my part. Filling out those forms might have taken three or four hours, total, over twenty-five years—about the length of *each* meeting of Grad Council![5]

The effort to change the rule on graduate committees was for a simple, easy, change. For policy changes that are more challenging, expect more work, more time, and more hassles. You have to want it—*you have to really, really, want it.*

One department that did want it that badly when I was Graduate Council chair was Chicano/Chicana Studies. They were creating the first U.S. Ph.D. program in the area. They'd already spent several years working with the administration to recruit new faculty and to develop their initial proposal. That proposal landed in council right as I was starting as chair. The initial proposal was lacking in a number of ways and it *terrified* me—I had nightmares of protesting students camped outside my office if we didn't endorse it.[6] To avoid that, I began a deliberate strategy of working closely with the department so that even if the initiative failed, I might dodge the blame. I spent a lot of time working with the team spearheading the proposal, helping them strengthen it into something that could gain approval by the Academic Legislature.

During that process, something interesting happened—I became a convert. What they were doing was important! UCSB is a Hispanic-Serving Institution,[7] and we should have a graduate program in Chicano/Chicana studies. But for the Department to launch the new Ph.D. program and admit their first student was a huge lift: hundreds and hundreds of hours spent by faculty, staff, and administrators in the department, within UCSB, and across UC over the better part of a decade. All that for an initiative that garnered broad support and encountered no active opposition—just the normal academic oversight involved in creating new programs.

It is possible to create structural and policy change in universities, and it is important to do so at times, but be prepared and committed to the route. Chicano/Chicana Studies was important. But, without energy, commitment, and a supportive team, initiatives easily die as their leaders are sucked dry and their withered husks are left to drift away. This may sound like a severe (and

---

[5] Worse, the next year I was appointed chair! I've since chaired other senate committees, and I'm now associate dean. Be careful of the roads you start down.

[6] That vision might have been deeply unfair, but the department's students had a well-earned reputation for activism.

[7] UCSB was the first HSI in the Association of American Universities.

common) criticism of universities and of their stodginess and obstructionism, but I don't mean it that way. Universities' core mission—creating new knowledge, teaching students, and producing the next generation of scholars—has remained steadfast across centuries and so we should not be mercurial in how we operate. Universities *should be* deliberative and thoughtful. We create the future, but we are grounded in the past. New programs should not be created hey-willy-nilly, because then they'd probably also end up being cast off, equally hey-willy-nilly.

Most policies make sense and target laudable goals. But there is a common phenomenon in institutions of all sorts: When someone does something bad, but which doesn't clearly violate any specific rule, there is pressure to create a new rule to prevent that bad thing from *ever* happening again. We can't rely on social common sense, what we might call "Grandma's Rules": Did you honestly think that was acceptable behavior in my house? So we create a new, enforceable "keep anyone from ever doing what Josh did" rule. Those new policies are often written broadly. They sometimes create new oversight or approval processes that make work for everyone. They probably don't even stop people from behaving badly—bad actors tend not to be rule followers.

I'll regrettably include, as an example of best intentions going awry, California's law on sexual harassment training, which mandates two hours of training every two years. I fully endorse training employees in what constitutes harassment, how to report it, and how the University deals with complaints. The in-person workshop I took when I arrived at UCSB was outstanding—it was informative, useful, insightful, and reassuring. But, to meet State law,[8] the training now is online and you can just let the timer run, periodically waking up to answer some questions that you don't even have to get right! Instead of being useful and insightful, it has become electronic box checking.

Of course, sometimes you run afoul of a problem where there ought to be policy, but there isn't. Staff members might be able to offer suggestions, but if there is no official answer, it's left up to you to do the right thing. Some years ago, I had a postdoc who got pregnant. UC had clear maternity leave policies for students, staff, and faculty. But postdocs? They fell into the cracks and only got six weeks of federally mandated unpaid leave. I put her back on the payroll full-time after that, even though we both knew it was a fiction. She wasn't going to be working full-time, but she needed to pay her rent and eat! We agreed that she'd figure out what she could accomplish and we'd sort it out between us—campus didn't ever need to know. But, a postdoc shouldn't have

---

[8] A 2004 California law requires that all supervisors receive two hours of training every two years. And "supervisor" catches almost everyone, even a graduate student who has an undergrad working with them.

to rely on their advisor's good will—there *should* have been policies. There are now. Postdocs have since unionized and are assured four weeks *paid* parental leave.

6.2 Practice

Many things regulate our professional lives that aren't *policy*, in terms of being written down as formal rules, but are rather institutional culture or "practice." That might be how we choose to translate policies into action, or just habits we've evolved. *It's how we do it here.* For example, with ten University of California campuses, a consolidated central administration, and a common set of systemwide rules and regulations, you might expect that campuses do things in the same ways. But we don't. Campuses developed their own ways of dealing with issues. For example, UC campuses have the same official salary scales, but most professors have off-scale salaries. How much off-scale, and how that is decided, varies across the campuses. At the extreme end of the power of practice over policy, consider Great Britain's unwritten "constitution," which is almost purely tradition and culture.

When ways of doing things are grounded in culture, they may be either easier *or* harder to change than if they are based on formal policy. It can be easier because there's no formal rule in the way. But it can be harder because culture is deeply seated. The very person who might have the greatest experience with the rules and so have the deepest insights on how to work around them may also be the senior staff member most embedded in institutional culture. "This is how we've always done it" or even "this is how it's done" can become potent arguments—even, or perhaps, especially, when they shouldn't be.

One thing to keep in mind is that humans like logic, order, and rationale: We will create logical-seeming narratives to explain things that are fundamentally illogical, nonlogical, or just lacking explanation. Consider conspiracy theories. Such created logic can take on a life of its own and become the official explanation, even when it's wrong.

For example, we heard once that campus had taken disciplinary action against a faculty member, but to protect the individual's privacy, it couldn't release any information about the case. The gossip and rumor mills started churning, and several stories burbled to the surface as the "truth." Whether any of them bore even a remote connection to the real truth was unknowable to most. Nature abhors a vacuum, and people equally abhor an "answer vacuum"; accepted "truth" can grow from pure fiction.

Another such case was my fault. When I arrived at UCSB, I was assigned to teach a course (*The Biological Environment*) in the Environmental Studies Program's introductory series. It was a Spring-quarter class, which ran until mid-June, and so I couldn't leave for field work in Alaska until early July—after soils had thawed and the associated flush of soil nutrients had passed. This became a real problem. I proposed restructuring the curriculum, to make the freshman series more overarching and integrative, and to follow it with sophomore classes that would build off that base. The first of these would be a Fall-quarter applied ecology class that I would teach. Bingo! With no undergrad teaching Spring quarter, I could catch the Arctic snowmelt![9] Now, twenty years later, we are revisiting our curriculum, but that Fall-quarter ecology class remains grounded in tradition and has acquired its own logic. Having established the structure, it becomes hard to change. It can be hard to know truly why we do things in our institutions, and whether the explanations we create for ourselves are about justifying our actions, or about evaluating from first principles why we do them.

One aspect of practice is how units communicate—or miscommunicate. Miscommunication is at the root of many problems with administrative systems. Kate didn't respond to Ed's paperwork because it looked like a duplicate of the request she'd approved last week, but Ed is waiting for Kate's response. Meanwhile, you're tearing your hair out because you need your student on the payroll! As long as Kate and Ed are just sending each other e-mail, which is rarely written or read carefully, the problem goes unresolved or becomes further muddled, until it explodes into a crisis that sucks in business officers, chairs, and deans. Yet, one two-minute phone call early on could have cut the clutter and resolved the confusion. But do units have a culture of "Just call Kate"? If they don't, and you find yourself in this situation, make that phone call happen.

6.3 Proprietary Systems (Technology)

Faculty sometimes, maybe often, feel that our systems should never block us from doing things that we want to do or that we think are academically desirable. That attitude, unfortunately, grows partly from a mix of ignorance and/or arrogance. More importantly, it isn't useful. The systems are what they are, and we have to operate within their constraints.

---

[9] It produced a stronger curriculum, but my prime motivator really was about not teaching Spring quarter.

I'm writing this in Microsoft Word, and if I try to tell it to make this text 11.1 point it gives me an error message: *"That is not a valid number."* I can yell at my computer that that isn't reasonable—if you can do 11.5, why not 11.1? But it doesn't listen—someone set the rules and that's just the way it is. There may be a good reason, or it may just be that Microsoft engineers didn't think we'd ever need 11.1 point type. Either way, that's how it works. It's the same with our university systems. It might be possible to get the system changed in time, but for now you have to live with it.

Technical systems age and get creaky, based on increasingly obsolete technology. So you either get the software update or you bite the bullet and get a whole new system. The University of California's human resources system, for example, had been based on 35-year-old code cobbled together with electronic duct tape and bailing wire that each campus had mutated to meet its own needs. UC decided that we'd reached the limit of patching and decided to create a new system. Streamlined and consolidated, it would save time and money. You can guess what comes next.

University human resources (HR) systems deal with a wild mix: Fall quarter, a graduate student might be on an externally funded fellowship; Winter they're a teaching assistant; Spring they're on a research grant. Through all this, they might do side work for their department, paid hourly. Developing an HR system was predictably way more difficult and expensive than predicted—it's been a nightmare. But, as campuses figured out the system's limitations, they developed approaches to limit the damage, such as issuing cash cards to students whose paycheck was delayed. We can hope solutions appear in future upgrades, and that it will be like breaking in a pair of shoes—they may cause blisters at first, but ultimately become comfortable and provide good support.

We depend on our systems, but they can be inflexible. When we run afoul of the limits of these systems, we quickly run into the old adage: what can't be cured, must be endured. But, although you probably cannot just push through the challenges, you can sometimes find ways around them. Well, likely *you* can't, but your staff, who work with the systems directly might—if they are interested in doing so for you. But that is the topic of the next chapter.

Clever staff can sometimes figure ways past either policies or systems. For example, if I really, *really* want 11.1 point type, maybe I could use 12 point, convert to an image and scale it down to 92.5%. That would be a hideous kludge, but maybe good enough—I've seen clever people come up with analogously ugly ways around policies. I like clean and elegant, but if it achieves the goal, I'll accept ugly.

6.4 People

People are the core of every human system. We create our systems, and even when we make good decisions, we still end up with policies, practices, and systems that may obstruct us. Those obstructions are worse, of course, when we've made bad decisions: bad policy, bad design choices, and so on. But sometimes the problem was a bad hiring decision: Systems fail because a particular individual gums up the works; they might be out-and-out incompetent or they might just be difficult and obstructionist.[10]

People problems can arise with several groups, and solutions differ depending on who it is:

1. A staff member: You are not their boss so you can't just order the problem out of the way. In fact, you aren't even in their "Chain of Command."
2. An administrator (chair, director, dean, etc.): They are *your* boss. So you *really* can't order the problem out of the way.
3. A colleague: They are your peer, possibly your senior. So again, you can't order the problem out of the way. This is the realm of politics.

Did you notice: you can't just order a problem out of the way with *any* of these groups? Getting pushy—giving orders—will therefore almost certainly fail, and likely backfire, damaging your standing without achieving your goal. The only people you can try to give orders to are those you supervise—students, postdocs, technicians—and your ability to tell even them what to do is constrained. In any case, problems with supervisees are problems of mentoring, not systems.

6.4.1 Obstructive Staff Members

A key here, as I noted, is that as an academic, you are not in the staff's chain of command. *You are not their boss.* Even when I was chair of the Environmental Studies Program, the only person I officially supervised was the departmental business officer; *she* supervised the rest of the staff. So, if a staff member is the problem, pushing will likely jam them in tighter, while getting mad will likely only jam them in tighter yet, making them less likely to want to help. Rather,

---

[10] That person might have been good once, but burned out and got sour, they might have hit their Peter Principle limit and been promoted to their level of incompetence, or they may just never have been very good.

when one person plugs the pipe, you need to either (a) find a way to "lubricate" the clog, (b) find a way around the person entirely, or (c) jump past it to the person who can clear it from the other side—their boss.

Working effectively with the staff is the focus of the next chapter, and so I won't spend much space here. The core motto of that chapter is "take care of the people who take care of you"; if you do that, building productive long-term relationships, you will minimize "system problems" that are really staff problems.

Sometimes, however, the problem really does lie with one person and you can't "lubricate" your way to a solution. Then you need an alternate approach route to your goal. Your car is broken? Take the bus. The road is closed? Take the train. Is there a way to sidestep the person? Is there someone else who can address the problem? Talk with other faculty who've had to deal with this person and might have solutions. Or to your departmental manager—the senior staff person in the department—he or she is well-connected on campus and might be the obstruction's boss.

Ultimately, lubrication and avoidance may sometimes both fail. Then you are forced into a more direct approach. In 2005, I spent the summer at the CNRS[11] in Montpellier, France. I'd arranged with my host, Stefan, that my wife Gwen (who provided Information Technology support at Santa Barbara City College) would need a desk and network access so she could telecommute. When we got to Montpellier, though, Stefan was battling with their head of IT, who had a bad reputation—"tin god" was an expression I heard. His response was, "Non, non, non, sécurité, pas possible, non, non . . . ." That went on for *two* days. We were freaking out because if Gwen couldn't get online, she couldn't work, and if she couldn't work, we couldn't stay—but we couldn't go home because we'd rented our house! Stefan finally went to the director of the institute, who came out of his office, grabbed Gwen's computer, stomped into the "tin god's" office, and slammed the door. There was some rumbling behind the door, and then the director came out, smiled apologetically at Gwen, said "c'etait tres simple, pas d'problème," and handed her laptop to one of the IT techs to deal with. That was the last of it; we had a wonderful summer and everyone else at CNRS was helpful and supportive. Banging heads with the wrong person was achieving nothing—even for Stefan who had gotten the arrangement approved and was a researcher at the institute. But once he went around the tin god to the tin god's boss? Then truly, c'etait tres simple; pas d'problème.

---

[11] Centre National de la Recherche Scientifique.

6.4.2 Administrators

University administrators generally come up through the academic track; with rare exception, we don't appoint deans or chancellors without doctorates![12] But we don't train for the leadership roles required to run a university. Good academic administrators are rare, and transitioning to administration (e.g., dean) is a choice, one few of us make and none are required to make. We might say "congratulations" to someone's face when they accept a position as dean, but behind their back... we wonder what they were smoking.

When we spend years developing our personal identity as scholars, leaving that path can seem like converting religions: you've abandoned the "True Faith!" We often see becoming an administrator as "selling out." Thus, we generally have a small administrative talent pool and leadership often suffers. But we depend on good people taking on administrative roles, so realizing that your interests are evolving to academic leadership is valid: How can you use what you've learned to advance the careers of others and to support the institution that has supported you? It can be a form of mentorship and of service to, in the words of Robert Baden-Powel, "leave this world a little better than you found it."

What do you do when a single administrator is the obstacle in the system? If the person is your immediate boss—chair or director—you're stuck. After all, they are your boss! That gives you three options: (1) shut up, put your head down, and get on with it; (2) start applying for new jobs; or (3) stage a coup and fire the bum. Most often the best advice is #1: learn to live with it.

I *strongly* advise against Option #3, because it can be suicidally stupid. I touched on my experience with this in Chapter 2, when I needed additional lab space. The impolitic words I said to the new University of Alaska Fairbanks (UAF) chancellor at her open house on "improving research productivity" were, "if you really want to know what you could do to enhance my research, it would be to fire my institute director." It was obvious to most that the director was at the end of his tenure and had little energy for leadership. Most faculty left him in his office and got on with their lives (Option #1), but because I needed his support to get additional space I couldn't take that approach, given that his inaction threatened to stymie my career. I didn't want to leave (Option #2), so all that was left was Option #3: go on the warpath. My foolishness (or desperation) helped catalyze the senior faculty to pressure the

---

[12] Janet Napolitano, former President of the University of California, had been a governor and a U.S. cabinet secretary but lacked academic leadership experience. But the challenges at Systemwide are managerial, rather than academic—running UC is about large-scale administration and politics.

director to retire. When I left UAF several years later, it was to be with Gwen, not out of pique. I remain personally connected and loyal to the Institute of Arctic Biology, and to UAF.[13]

Mostly though, if the problem is your immediate boss I would recommend either Option 1 or Option 2: put up with it, or leave. The effort and time it can take to try to stage a coup might well destroy your career even if you are successful—wars leave casualties strewn about. If a coup is something that really should happen, why haven't the senior, *tenured*, faculty already pushed it? Being able to safely act in political messes is a major reason tenure exists. Assistant professors should not have to lead in political battles! But senior faculty also have a longer perspective, are likely have the resources they need, and are more able to weather a crisis; they may feel less inclined to battle. At UC, department chairs rotate every three years—let us hope that by the time you realize someone is a disaster, their term is about up.

If the problem person is further up the chain of command—dean, provost, and so on, it probably means that your institution is more generally a managerial mess, so let it go. Or leave. It would be rare that someone more than one step up the chain of command could have enough direct influence to make your life a misery.

6.5 Politics

Two key definitions of "politics" in the *Oxford English Dictionary* (OED) are: "Actions concerned with the acquisition or exercise of power, status, or authority" and "Management or control of private affairs and interests, esp. as regards status or position." Both definitions highlight *status*—institutional politics are about the human interactions to manage affairs, with an emphasis on your position within the structure.

Academic politics have a terrible reputation, the classic quote being "Academic politics are so vicious because the stakes are so small," although who this should be credited to remains unclear—maybe we could fight over it? But all the way back in 1765 Samuel Johnson said something that captured that sentiment and explained some of the mechanism:

---

[13] I remain particularly grateful to my colleague Dr. Gerry Shields who backed me up at that event with the chancellor, and to Dr. Bob White, who took over as director and was wonderfully supportive.

> But whether it be, that small things make mean men proud, and vanity catches small occasions; or that all contrariety of opinion, even in those that can defend it no longer, makes proud men angry.

University politics at multiple levels can cause trouble—within your department, between your department and others on campus, and between higher-level administrative units. Politics becomes a problem when people or units are competing for control over a decision—you're caught in a turf fight, or a pissing match, instead of hung up by a rule or policy.

Developing initiatives within a department is inherently political, as you need buy-in from colleagues. Will the next hire be in your area? Will the department prioritize your initiative? This is classical political turf, involving personal influence to muster support among peers. Although logic and argumentation are important, often relationships and personal calculus can be more so. What will supporting *you* gain or cost *me*? Will it buy support for my initiatives? Will it block my plans? Maneuvering these might call for evaluating which groups within your department it would be most beneficial to affiliate with and to support.

Departmental politics are usually not immediately "transactional" (i.e., do something for me and I'll do something for you) but are more likely grounded in competing visions—should the department build in popular areas, craft a unique identity, or build clusters of excellence? The power of your voice in these discussions will likely depend on your status within the department. Status depends on factors including time in the program, personal charisma, and credit earned through departmental service. External professional status also matters—if your colleagues decide they made a mistake in hiring or tenuring you, you'll likely be a political outcast. If on the other hand, you achieve recognition that brings campus kudos—an National Science Foundation CAREER grant or a prestigious award for your first book—it will enhance your departmental stature. Greater standing will make it easier to convince your colleagues to support your initiatives.

Healthy departments operate on the appearance that arguments are won by logic rather than power—largely because there isn't much basis for direct power—I have my lab, my funding comes from outside, and so on. A colleague could threaten to vote against a merit raise, but it's hard to get away with overt thuggishness. Often, the less narrowly self-serving an initiative appears to be, the more likely it is to succeed. We may look at proposals through the lens of "How will this benefit me? But we aren't supposed to act on that basis. Political dealmaking is either more subtle or much more open and enshrined in a department's planning and vision. For example: "Priority #1 is the evolution position, but #2 is the plant ecologist." It's still all politics.

6.5.1 Staff Politics

Faculty–faculty interactions are inherently political, but is that also true of faculty interactions with administrative and departmental staff? No. The staff are my teammates, and I hope my friends; I work with them, I pay close attention to their perspectives, and may sometimes debate with them over how to achieve particular goals, but I don't *play politics* with them. We operate in different arenas. For example, Eric Zimmerman has managed undergraduate affairs (advising, class scheduling, etc.) within the Environmental Studies Program since I joined UCSB. But faculty are responsible for the content of our curriculum—what our students *should* learn. Eric's job focuses on the logistics of making the curriculum happen and on supporting students as they move through that curriculum. He has deep insight into student perspectives, and I would never dream of proposing curricular changes without consulting him—I try to avoid being that stupid. But if our interactions have become *political*—that is, grounded in acquiring or exercising power or status—something has gone seriously awry.

Politics certainly operates at higher levels, between administrative units—does Academic Personnel have the last word on something or does the dean's office? Those may be fought out among the senior staff or unit directors. Such higher-level battles can control how operations work, what the rules are, and who you have to talk to in order to get things done. Universities are rife with such politics. Those debates can be generated by any of a number of causes, including new laws or policies flowing from higher up in the system. It may be unclear how to address those new mandates or even who is responsible. Mostly such battles define the institutional structure that you have to navigate—who do I talk to? But you aren't likely to change the policy landscape that caused the political battle. Best not to waste your time trying—you could, instead, be writing your book or teaching your class. Avoid those higher-level battles until you are established, if then. Especially avoid battles involving units that don't embrace the missions of teaching and research. In my experience, physical plant and maintenance units are most commonly guilty on that front—for example, at an academic senate budget and planning meeting about funding building maintenance and recharge, when a committee member said, "We're all on the same team," the head of physical plant responded, "I don't see it that way."

6.6 Pecuniary Realities: Money and Finance

There are few universal truths, but one of them is unquestionably, *There Ain't No Such Thing As A Free Lunch*. Everything we do costs money and likely more than we realize—universities never have enough money. To develop a new initiative, we always need to figure out how to pay for it, or perhaps more realistically, who should pay for it. As we contemplate finding funding, one of the most useful pieces of advice I've ever heard was a little parable from UCSB's provost of the College of Letters and Sciences. And no, it isn't, "the best things in life are free." Nor is it, "money can't buy me love!"

Rather, the provost said, "When I was a professor, I was sure my department chair had some secret stash of cash he was holding on to. But, then I became chair, and I was sure that the dean had a secret stash. Then I became dean, and I figured that the provost must have a secret stash. Now I'm provost, and so I'm sure it's the vice chancellor who has a secret stash!"

To take that further, almost everyone in the University of California system takes it as an act of faith that the president of the UC system has a secret stash in *their* budget—how else could the systemwide Office of the President (UCOP) have a budget almost the size of one of the UC campuses? Mustn't there be waste, inefficiency, and hidden stashes? If UCOP were just more efficient, couldn't they free up millions for my pet initiatives?

Well, no. When I served on the UC systemwide Committee on Planning and Budget (UCPB), I learned that not even the president had a secret stash. Most of the UC Budget was pass-through funds to programs like Education Abroad, the Division of Agriculture and Natural Resources, and others. In its administrative operations, even an external audit found that UCOP was actually fairly lean. Huh! I also learned how complex UC really is, and how much talent, skill, knowledge, and dedication it takes to keep us functioning—you don't manage a 250,000 member retirement program with over $25 billion in assets by hiring a few graduate assistants.

The "No Secret Stashes" rule is universal. Certainly it's true at all public universities—maybe Harvard has "secret stashes," but I suspect Harvard's leaders keep looking for them, too. Every good administrator tries to keep some reserve to cover crises and opportunities, and the further up the system you go, the larger the absolute size of those reserves may grow. But they are rarely, perhaps never, big enough. And crises and opportunities do happen.

The bottom line is that universities are wildly complex enterprises, which have developed complex systems—financial, administrative, and institutional—to allow them to operate. Some of those systems are awkward, inefficient, clumsy, and you can probably add a few more snarky adjectives to that list. But there was, once upon a time, a decent reason for developing them. Regardless of the when's, how's and why's, though, *they are what they are*. If you want to live and function within the system, and particularly if you want to try to make changes, you need to understand the systems. And hopefully, to have some sympathy for them, and for the people whose job it is to keep them functioning.

As faculty members, we don't need to become experts in our administrative systems—that isn't our job and it isn't our interest. But understanding the overall structure, and some of its logic, will help you manage your career, and is vital if you want to make programmatic change.

7
Working with the Staff

Take care of the people who take care of you

As faculty, we recognize that we carry the mission of the university—we lead the research and we teach the classes. Ultimately, the entire university is about us. Everyone else in the system—from the President and Provost in their corner offices, to the painters and plumbers in the maintenance shop—exist to enable us to do our jobs.

As academics, therefore, we're like fighter pilots—the success of the entire operation ultimately comes down to our ability to perform. We've been carefully selected and trained through years of the most rigorous and intellectually grueling training our institutions can develop. We're the best of the best. Top Gun, or maybe Top Pen. Is it any surprise, therefore, that some faculty also develop the arrogance that is the stereotype of the fighter jock?

But every fighter pilot knows, when they climb into the cockpit, that they aren't going anywhere unless their maintenance crew has first fixed, fueled, and armed their jet! They know that if their ground crew was careless, they could end up a greasy smear in a smoking, flaming hole in the ground. For a pilot, there may be no more important people in the world than their ground crew. The mission's success may ultimately rest with the pilot's ability to deliver, but it takes a team to put them in the right place with the right tools.

As academics, we're no different, although it's only our careers and not our actual lives that are on the line. We may carry the mission of the university, but we won't carry it very far without our team behind us. I don't prepare my paycheck. I don't maintain my students' records. I don't "hit the button" to submit a research proposal. And I don't fix the plumbing in my lab when it breaks. I rely on others to do those things, and I rely on skilled administrators to organize the whole, making sure the right people are doing the right things in the right places at the right times. Without the administration and operating staff, I'd be sitting in an empty field pontificating like Plato.

But I'm not sitting in a field pontificating like Plato. I'm sitting in my office in a building on a campus with >23,000 students and $180 million a year in research. University of California Santa Barbara (UCSB) has ~1,000 faculty members, but to enable us to do our jobs, it has over 3,500 administrative and staff employees; yet even with that number, we are understaffed.

Great staff is part of what makes a great university. We rely on knowledgeable and skilled professionals, and as systems grow more complex, that need increases. Together, faculty and staff form a single team: we have different responsibilities, but make shared contributions. When the staff are good, I can focus on my job. Yet, they work in an environment where, all too often, the motto might be "Perfection is invisible; anything less, complainable." No one appreciates that. Understanding your staff is key to working effectively and to focusing on your job. They are our team members, our colleagues, and we hope, our friends.

Although faculty and staff together make a team, we play different roles and we come to our jobs differently. Our staff aren't us. If someone asks me who or what I am, I'll almost certainly say, "I'm a soil ecologist," or, "I'm a professor." Being a professor isn't just my job, it's my identity. I am a professor. Although many staff members are as dedicated to the university and to our mission as I am, I suspect few would self-identify based on their job title. For most staff, their jobs are what they do, not who they are.

Staff members don't work for us individually—they aren't my staff. They work for the department or for the university. In my department, I am only one of thirty-plus faculty who need support. My needs may be important, but no more so than those of others in the department. I can't expect that my grants coordinator will be able to drop everything to help me out; someone else also may have a proposal that needs to go out this afternoon. And, of course, staff members' workday, unlike ours, officially ends at 5 PM.[1]

Lastly, keep in mind, as well, that many of the people who work in our institutions do so because they need a job. Hence, the spread of skills, drive, and ability can be wide. Our best staff are awesome, but not all are that good, and a few aren't very good at all. Some problems arise because a staffer is new and still learning their job; a good supervisor will catch most of those errors. But some people continue to make mistakes, either because they are poorly trained or are in a job they are not suited to—this reflects a failure on their supervisor's part. Being in a well-run department, where staff operations are efficient, competent, and supportive makes a huge difference—you get to focus on your job and rely on the staff to support you in doing so. Not everyplace, however, runs smoothly; some programs' paperwork is almost certain to be a mess.

---

[1] Though, of course, it often doesn't.

7.1 How to Get Stuff Done

I've been at UCSB for over twenty-five years, and during that time, I've figured out how to do many of the things I need to do, which largely consists of figuring out the appropriate paperwork: how to assign grades, order research supplies, arrange travel, and so on. Yet, I still regularly run into things I don't know how to do. Sometimes that's because I haven't had to do it before, sometimes because systems have changed, and sometimes because I just forgot. It doesn't matter, since the answer is always the same, as noted by Stephan Heard:[2] *Find somebody to ask.* Or perhaps, I can riff off his comment and adapt Jefferson Airplane:

> *Don't you want somebody to ask*
> *Don't you need somebody to ask*
> *Wouldn't you love somebody to ask*
> ***You better find somebody to ask!***[3]

But who is that somebody? Universities may all do the same things, but each has evolved on its own path over decades or even centuries—each has its own unique ways of doing things. Even within the University of California, each campus does things differently; knowing how "UC works" could lead you astray at a different campus. So Principle #1 is find somebody to ask!

Most questions fall into broad staff categories: You have a question about budgeting for a proposal? Talk to the contracts and grants person. A question about graduate students, talk to the graduate advisor. Paperwork for your promotion case? Talk to the academic personnel coordinator. If it isn't obvious who the right person is, try your department's business officer/manager—if they don't know the answer, they know who does (or at the very least, who knows who knows the answer). It will almost never take more than a few "talk to so-and-so's" before you can get any question answered. Have patience. Department managers typically have years of experience on campus, having reached their current position by advancing through the staff ranks, starting with low-level positions in small departments and getting promoted to more and more senior positions. In getting there, they come to know everyone who is anyone, establishing networks of friends and relations among

---

[2] Stephan Heard is the author of The Scientist's Guide to Writing, Princeton University Press, and he offered this pearl while I was exchanging e-mail with him about my ideas for this book. Thank you Stephan!
[3] Of course, now I have Grace Slick roaring in my ears.

the "lifers"—the senior staff across campus. These are the people who really run the university.

Learning who to ask is a good start, but as you adapt to your new habitat and learn the lay of the land, to work most efficiently and to get the best out of your team, you need to go beyond our Principle #1. You should develop some understanding of what each staff member does, and of the constraints they operate under, so that you can coordinate with them effectively.

When I was chair of the Environmental Studies Program, I never learned all of what kept Cheryl busy all day long managing the program's finances. I didn't need to—my job wasn't to be able to replace her, but to support her in her job so she could support me in mine. I never had to get hands on with our financial systems, but I had to know about the various funds and "flavors" of money—how we could use donor funds in different ways from State funds, and how Temporary-Sub-0 was different from Permanent-Sub-0.[4] I needed to know enough that I could work with Cheryl to try to solve problems creatively. And, I needed to accept that if she said she couldn't do something, it couldn't be done. I knew that I could trust and rely on her, knowing that she would deal with things that needed to be dealt with, and would only come to me on issues that needed the chair's decision.

As I've migrated up the ranks, the importance of the staff in supporting me and my mission have become increasingly apparent. As a graduate student, I mostly dealt with Fred in the stockroom who ordered stuff; he helped me survive grad school. When I was working late in the lab, I'd take breaks and hang out with Dorothy "After 5, I'm the Boss" Brownlee, who ran the Hilgard Hall custodial team; I adored her—she could lighten up any evening. After I became a professor, I interacted with a larger group of staff who managed a larger network of activities: the grants manager who helped me figure how to juggle money between projects, the grad advisor who helped sort out student funding. Now I'm an associate dean. When I'm responsible for some faculty member's merit review case, I rely on Shawnee having meticulously gone over all the materials to make it ready for me to "fly." She is my crew chief, but instead of it being my life on the line, it's someone else's—maybe yours.

So how does this translate into guidelines for working with your staff? My version of the Golden Rule is take care of the people who take care of you. That translates into a number of simple guidelines, guidelines that reflect that our

---

[4] Don't ask.

staff are friends and team members, but they are not academics. The best staff are as good at their jobs as we are at ours—and I can't do theirs.

When you are working with the staff, who is "in charge" can dance around. When I chaired Environmental Studies, the program's organization chart showed that Jo Little, the business officer, reported to me. But I always called her my boss—she ran the shop and she knew how operations ran more deeply than I ever would (or wanted to). We made a good team—I followed her lead on administrative matters, I led on academic ones. Now, as associate dean, I have even more "bosses"—Alex, Heather, Shawnee, and Kathy each rule in their own territory. My real boss, the dean, operates similarly: He doesn't make decisions without input and advice from the experts on his team.

7.2 Building Bridges

Getting along with our staff teammates usually isn't hard. Just being friendly and respectful, and acknowledging that they probably do know their jobs is a start. Most of the staff I know are good people who I enjoy hanging out with, more so than with some, maybe many, of my professorial colleagues. So step one is to just be nice and avoid going into "fighter jock" mode—nobody likes a jerk.

Beyond that, people bond over food. To build his team, Pierre, the dean, takes us out for breakfast to celebrate birthdays—it's an opportunity to get together off campus but without imposing on people's personal time. In Environmental Studies, our business officer organizes happy-hour get togethers. I rely on my "un-secret weapon"—my wife, Gwen. She's a great baker, and likes experimenting with new recipes, but her normal rule is, "get it out of the house" because otherwise she'll eat it! Those cakes go into the office break room: Gwen gets the cake out of sight, coworkers get treats, and Josh gets a little goodwill—everyone wins.

It usually only takes a little effort to build bridges and lubricate the system. The occasional nice gesture (such as leaving Gwen's baking in the break room) earns "brownie points." When I travel, I might also bring back something for the break room—for example, chocolate from Switzerland. If you try to "lubricate" at the moment you need help, however, it looks like bribery, and may backfire: "You only pay me any attention when you need me to stay late, and you think a chocolate bar is payment?" Of course, a small thank you gift after the fact for a staff member who really went above-and-beyond to help with something is fine—then it's a true thank you. Keep your systems running by regularly being friendly and nice. Take the time to occasionally "check the oil" even, or especially, when the "engine malfunction" light is not on.

Why is it so valuable to invest a little time to build those bridges? Beyond, of course, just making life more pleasant and fun for everyone? As long as all we need is for our staff to do the simple and direct, straight-and-narrow, of their jobs, most staff will be professional even if you treat them poorly. But what about when you need to figure a way around the rules? Some solutions may be a little dicey, bending rules to the point they begin to groan and crack. Would you risk your job for someone who treats you like crap? Don't expect someone else to!

When I moved to the University of Alaska Fairbanks, I had a long cold drive from Michigan in January. I stopped in Montana for a day to visit a friend. But after I arrived in Fairbanks, my travel expenses bounced. The finance person in the institute told me the campus auditors questioned why I'd spent two nights at a motel in Montana; if I didn't have a good explanation, they would reject the cost of the second night. I told her the real story, and then we stared at each other, hemming and hawing. Then I had an inspiration: "just say there was snow in the mountains." She looked at me for a moment, slightly puzzled, and then a smile spread across her face. "I love it" she said. We both knew that central accounting just needed a rationale for the paperwork to justify the extra night; I guessed that they would interpret my simple, and absolutely true, statement to imply that I couldn't get through, rather than that I spent the day skiing with a friend. I was right—that was the last I ever heard of it.

Years later, after I had moved to UCSB, a student of mine spent the summer working at the Toolik Field Station in arctic Alaska. His project involved seemingly infinite hours of field and lab work. He'd initially purchased a round-trip ticket to span the entire summer, but then a friend decided to get married mid-summer. For my student to attend, the cheapest way was to buy a new round-trip ticket from Fairbanks to Seattle rather than changing the ticket he already had. The question arose: why did he fly out of Alaska? Why should the grant pay for apparently personal travel? Fair questions, but had we bought two separate tickets—up and back for the first half of the summer, and then back up later, after the wedding, the question would never have arisen. The issue was optics, not ethics.[5]

---

[5] I'd budgeted multiple trips, and I flew up twice that summer for a week each time. No one should be expected to spend three months at the Station working eighty or more hours a week, while seeing if you can break the record for most mosquitoes killed with a single swat (~200) without getting out for a break.

So when our travel person asked about the student's trip, I offered, "He had to go to a meeting" (after all, a wedding is a "meeting"), trying again the technique of giving minimal, but honest, information that I expected would be misinterpreted. It didn't work, damnit. She asked "What conference did he go to?" Oops. I responded uncomfortably "I didn't say he went to a conference— I said he went to a meeting." She looked thoroughly confused. So I had to provide the full story, which made her complicit in my shenanigans. I will never have any ethical qualms about covering the trip off the grant, but I felt guilty about putting her on the spot to come up with some rationale that fit the rules to justify the trip. But I still needed her to do it—and that required having built a relationship of support and trust.

What do you do if you realize that you have burned bridges—that you've behaved badly enough to become persona non grata among staff units whose support you need? The best advice might be to start looking for a new job, where you could start with a clean slate (that I would hope you wouldn't mess up like the first), but that is probably impractical. My next best advice would be to apologize to people you've treated badly. Apologize, and promise to be better in the future—and then *be better*! I'd also recommend a chat with the department manager to express your regrets and your wish to change your patterns. Ask for their advice and help, and then earn it.

7.3 Problems with Staff Members

As faculty, we often have things we want to do, and sometimes those run afoul of the system. Sometimes it appears that our problem has to do with a particular staff member. Sometimes, that's true—a particular staff person may be the problem—but other times, they may only be the *apparent* problem. I identify four types of staff "problems":

Incompetent staff
Prickly personalities
Unreasonable requests
Bearers of bad news

These categories span from the real problem being *the staff person* to the real problem probably being *you*. And of course, they may overlap—someone who is incompetent might still be just passing along bad, but accurate, news. But how are you to tell whether their bad news is true when you have good reason to be skeptical of the information's source?

7.3.1 Incompetent Staff

If there is a staff member who keeps screwing up, the problem really lies with their supervisor. Is the employee not properly trained? Are they in a position for which they are not qualified?

It's natural to get angry at people whose incompetence is screwing you up. But how you express that anger is a choice—do you vent it at this person? Anger can be a useful tool, but you have to have the right target in your sights: the person who caused the problem, and who might be mobilized by your anger to fix it. But how likely is that? If this person is truly out of their depth, being angry isn't going to make them better at their job. If they are trying to fix the problem, venting may make you feel better, but it distracts them, slows them down, and makes them less likely to want to help you.[6] And they won't forget. Everyone loses. Brilliant!

Your anger also may merely lead them to a solution that deflects you, but which doesn't help you: "OK, I'll send this paperwork forward as it is, since you seem sure you know more about this than I do . . . " But they are thinking: "I'll let the other office bounce it back—that may waste everyone's time, but it gets you off my back."

Even if they did screw up, but acknowledged it and are fixing it, why be mad at them? We all screw up. All we can do afterward is try to fix it. If you have chronic problems with this person, talk to their supervisor or your department chair. Maybe the supervisor can better train or manage the person; maybe they can adjust the job responsibilities to better align their abilities with their responsibilities; maybe they can fire the person—though accumulating the documentation required to fire someone can be time consuming. In any case, that's not *your* job.

7.3.2 Prickly Personalities

If you have staff who are sometimes prickly, avoid anger—instead, try lubrication. It's more successful in both the short- and long-term. Lubricating your way through a problem involves sweet talk or sweet treats, and time. If you only have the time of day for the staff when you have a crisis, you're not giving them much reason to want to go to bat for you. To re-use my car analogy, you

---

[6] The frustration and impatience may be unavoidable, but find a different outlet than the person in front of you. Bitch to your partner at home, go running, take the dog for a walk, or yell at the wall of your office if you need to.

can't wait to change the oil until the engine seizes; it needs regular maintenance. People are the same. My department's staff are on the fourth floor of a different building than my office; I still make sure to go up at least every few weeks to say hi—particularly when I don't have anything specific to ask. They are nice people, and I value the relationships.

Some people might be prickly because they've been burned out by too many years of faculty swamping them with unnecessary last-minute crises. Sometimes, there's just cultural friction; I knew one person who had a reputation for being awesomely knowledgeable, but prickly; many people avoided her when they could. Mostly that was just her New Yorker coming through; but I grew up in Manhattan and we got along great, bonding over our shared home town.

The best solutions to dealing with prickly people are twofold: First, put up with them. If the person is capable but prickly, the important message is that they are capable. Great. The second is to try to build a better bridge, rather than merely minimizing your interaction with this person. Try lubrication, and see whether you can improve the relationship.

7.3.3 Unreasonable Requests: Expecting Too Much

Maybe because our lives aren't literally in the hands of our crew (as they are for fighter pilots), academics sometimes develop a bad habit of taking the contributions of the staff for granted. For example, when I was serving on a search committee for a grants coordinator, a faculty member criticized one of the candidates: "Yes, she's willing to stay late and put in the extra time when a proposal crisis comes along, but she lets you know she's doing you a favor." But—we were talking about a staff member who has a forty-hour a week job without overtime! If she stays late to get my proposal out the door, she is doing me a favor. It seems just common human decency to recognize it. Absolutely, I want staff who are open to doing that favor when needed, but they don't have to—they do so because they are committed to their job. They didn't sign on for our job, and expecting them to be on 24/7 the way we often are is beyond unreasonable.

Don't be the person who always shows up at 4 PM on the day of a deadline needing to get a proposal out the door. It will happen occasionally, and then you will have to depend on staff to dig you out of your hole. University staff know us, so they know the drill—but every time you get something in early so they have time to deal with it peaceably, you earn a little credit toward the occasional crisis.

My wife, Gwen, used to work in computer support at the California School of Professional Psychology. Once a faculty member needed a big desktop publishing project done that day. But, he didn't just dump it on her and split—he stayed at her office all day to be available if needed, and he brought her lunch. She'd have done that project anyhow, but she really appreciated how he supported her.[7] When you have a real crisis, be that person.

Some faculty "problems" with staff are really just with us and our unreasonable expectations. You can minimize those problems by adjusting your expectations to be less unreasonable, and by backing your team up when you do have something unreasonable that you still need to ask!

7.4 Bearers of Bad News

Most bad news comes from a staff member who is telling you that there is a problem: you have to do something a particular way or that you can't do it the way you wanted to. Most often, when they say there is a problem, there is, in fact, a problem—and it is not them. Rather it is likely related to policy or systems. The news might be infuriating, but this person is merely its bearer, not its source. Back in 'Olde England, harming a Town Crier—the bearer of bad news—was treasonous; attacking the King's Man was attacking the King. Attacking the dean's person...

Unless you have rock-solid evidence that the staffer is the problem, assume they are not. Rather, they are likely the person who might help you solve it. Don't get mad at them, get mad with them—at the stupidity of the system. They may find it just as annoying as you do—or may be able to explain why it isn't actually stupid, and that there is a good reason for the policy. Take your righteous anger, and instead of venting it at this person, turn it into motivation to change the rules.[8] The heat of your anger is almost always better used to fuel your inner fire, rather than to scorch someone else.

If the person you are dealing with didn't cause the problem and is just the bearer of bad news, getting mad at them is pointless and verges on evil: abusing someone who is innocent and who can't fight back.

---

[7] Enough to tell me the story thirty years later!
[8] As I did over Ph.D. committee composition as I described in the last chapter, but keep in mind the price I paid.

7.5 Imperfect Faculty: Kiss up—kick down

The previous sections focused on dealing with staff "problems," but sometimes the problem lies with us—the faculty. We can be incompetent when we deal with administrative systems and the people who maintain them. We, too, can have prickly personalities. And we can be impatient when our personal projects and initiatives run afoul of administrative rocks and shoals, even when the problem isn't the staff but our poor navigation—why didn't we take their good advice? It's not fair to get advice from someone on how to avoid the rocks, but to ignore it, and then blame them when you run aground.

There are faculty whose reputation in the dean's office is, "they're not our favorite person over here." Some of those names are obvious, because they are well-known challenging personalities. But other names have surprised me, because they are people I'd thought well of. Those people generally seemed to fit a more insidious personality type: the Kiss up—Kick down type.

Smart and self-confident, these people know they need colleagues to support their initiatives. But staff—they only exist to support the faculty and shouldn't get in the way. Treating people that way, though, isn't just being a jerk, it's being an arrogant, manipulative jerk. You may know people like this, but might not even realize it because they're kissing up to you!

If you are early in your career, and find yourself inclined to blame or to be rude to a staff member (or teaching assistant, etc.), particularly about issues that aren't their fault, break yourself of the habit. If you can't behave decently because it's the decent thing to do, do it because it's just good tactics. That person you're writing a nasty e-mail to, or about, is part, not just of your "ground crew," but of ours as well—and *we* care. You may keep your colleagues in the dark for a while about how you've been treating our staff team, but we will learn eventually—and it's not smart to be "not our favorite person" in the chair's or dean's office!

I realize that this advice is probably pointless—the people who need it most, probably won't heed it, and if you don't need it, you don't need it. But I'll still play Don Quixote and tilt at this windmill—please, please, please, don't be one of these people.

Take care of the people who take care of you—our systems work best when we treat our staff as the vital team members they are. Not only can we achieve our goals more smoothly and efficiently, but life is also more pleasant. Treating our support staff badly undermines all of us—it creates lose–lose situations that hurt productivity and make solving problems harder. Such behavior makes messes for your colleagues to clean up. It is, as was said of Napoleon's execution of the Duc D'Enghien: "C'est pire qu'un crime, c'est une faute." *It is worse than a crime, it is a mistake.*

SECTION 3
THE NEXT GENERATION

Those who can do, also teach

Nothing we do as academics is more important than teaching and guiding our trainees. We must "reproduce the species"—creating the next generation of academics. It's also important because in some areas of research, graduate students are in the lab or field doing the research that builds their advisor's program. But most undergraduates don't go on to graduate school—they go out into the "real world." We teach our students to be knowledgeable, skilled, critically thinking citizens. The challenges associated with teaching and mentoring are, however, multiple. We work with our trainees to help them develop, as well as to help them work through their personal challenges to get to the right place for them. We teach classes and have to incorporate that teaching into our overall academic programs and our lives. How do we become successful teachers, while still fulfilling our other responsibilities? Teaching and mentoring issues are likely the most challenging issues that we face as academics.

8
Mentoring 1
Vision and Philosophy

We make a living by what we get, we make a life by what we give.
Attributed, though apparently falsely, to Winston Churchill

The delicate balance of mentoring someone is not creating them in your own image, but giving them the opportunity to create themselves.
Steven Spielberg

Our job as academics is to create the future. We create knowledge, understanding, and insight; some of us invent things that transform our world. But the most important part of that future, and our most lasting legacy, is the people we teach and train. Our job is to help them reach their potential and become the best and most capable people they can be. We achieve that through mentoring.

The *Oxford English Dictionary* definition of *mentor* is:

> a person who acts as guide and adviser to another person, esp. one who is younger and less experienced ... an experienced and trusted counsellor or friend; a patron, a sponsor.

But that definition encompasses a wide and amorphous range of roles and a wide array of relationships, relationships that ultimately extend to everyone in your academic community. Mentoring may look different when you're talking to an undergraduate, a Ph.D. student, or a junior colleague, but the essence is the same—helping people get to the place that is right for *them*.

We take on mentor roles early—often as graduate students and certainly as postdocs. In my first year as a Ph.D. student, the senior students in the department provided critical guidance—including not to worry about getting a D on the midterm in soil physics![1] My most important mentor, though, in some

---
[1] Sam Traina warned me that it was common; the material would "click," and I should still get an A in the class. He was spot on, but without such peer mentorship, I'd have panicked!

ways, was Ken Killham. Ken was a postdoc in the lab, and when I wasn't sure whether an idea was clever or just stupid, he was my frontline, go-to, guide. I could run ideas past him without fearing the embarrassment of bringing a dumb idea to Mary Firestone, my Ph.D. advisor. I relied on Mary for big-picture stuff—and I still do—but I valued having a sounding board who was a senior peer. Later, I found myself the senior student in the lab and new students used me in that role. These informal mentorship roles are important, and they build mentorship as a culture—it's a state of mind, or even a habit, rather than a specific activity that takes place in a specific setting.

As we gain in seniority, even random conversations can take on a mentoring role. This was highlighted for me when I read a column Emily Bernhardt (2016) wrote titled, *Being Kind*, which was her President's Address for the Society of Freshwater Science. In her column, Emily mentioned several incidents that, for her, were notable acts of kindness when senior colleagues took the time to help her work through challenges. One of the people she mentioned was me. That surprised me, because although I remembered meeting Emily, my memory of the event was that I was having fun—meeting a talented and nice new colleague who was doing fascinating science. We had a great discussion, puzzling out an interesting problem; I certainly had no perception that I was being kind.

But that's the important point: What we experience as the motivation for our actions and what the recipient experiences as their outcome can be very different. A random, casual conversation can have an outsized influence. My former student Jay Gulledge once highlighted this, noting that one of the most significant events of his graduate career was an eight-hour-long "argument" we had about his experimental design on a drive from Fairbanks to the Toolik Field Station in arctic Alaska. In the end, I gave in and agreed; I told him, "do it your way." I remembered the trip, but hadn't realized that for Jay, it was better than passing his qualifying exams. He knew that I hadn't caved out of exhaustion—I'm probably more stubborn than Jay—but because I finally accepted that his approach was as likely as mine to succeed (and it did). For me, that drive was just a trip to Toolik; for Jay, it was a rite of passage.

An interaction at the first conference I attended as a Ph.D. student stands in stark contrast. I was chatting with Mary Firestone when a senior member of the field came up and said, "Mary, I see you're presenting a paper on heterotrophic nitrification." Mary introduced me and noted that I was presenting the talk. This person gave me a sideways glance, turned back to Mary, and asked "Do you believe your data?" I got the distinct impression that I was beneath notice. Perhaps I was overly sensitive and misread the intent, but I have never forgotten, nor forgiven, that interaction.

Essentially every interaction we have with students, postdocs, and junior colleagues becomes a potential act of mentorship and guidance, whether we like it or not. We are always "on," and for our advisees, we set a standard for what it means to be an academic—positively or negatively. We build a reputation, and such reputations get around. If you want to recruit top talent, and to get the most out of your people, treating them well is a good first step.

8.1 What is Mentoring?

Mentoring involves many dimensions, as mentees we look to a range of people for support (Table 2.1). As mentors, therefore, we can expect to only fulfil a portion of anyone's needs. We can only do what we can. Emily Bernhardt, in her column, emphasized the importance, and power, of being kind—supporting people, building relationships, and generally being a good person. At some level, her argument is rather biblical in its analogy to casting bread upon the water. Ultimately, it pays back by making a more productive environment for us all. One key point she made though, was to distinguish being *kind* from being *nice*:

> In using the word *kind* I very explicitly do not intend the sometimes synonym *nice*. As intellectuals struggling to understand the world around us it is vital that we argue, that we hone our understanding through challenging our own views and the views of others. We cannot, and should not, always be nice while intellectually sparring. Yet we can spar while still being kind. We can disagree with a point while respecting the person making it.

That, I hope, characterizes the eight-hour argument I had with Jay on the road to Toolik and why it was so important to him. We were banging heads, yes, but we were banging heads *together*, trying to come up with the best way to solve a problem. Jay recognized that he had grown past merely being my student, he had become my colleague and my peer—and that we both understood it.

Emily's argument is analogous to my motto: "friends don't let friends publish bullshit." But, where I push the message by being deliberately crude and a little shocking, Emily was expanding on the core and doing so both kindly and nicely—illustrating her message in her language. You don't serve your friends by giving them a free pass, but by helping them improve their work; this includes criticism. But professional criticism is distinct from personal attack. We've all had boneheaded ideas—that doesn't make us boneheads.

Many books have been written about mentoring and it's various dimensions—and by people who have thought about the issue in a more organized way than I ever have (e.g., Rockquemore 2016). For me, the essence remains my argument that your job is to help people get to the place that is right for them.

But how do you figure out what is right for any particular person? They may not even know! The obvious answer is listen. What do they say about their vision and wishes? Have that conversation early and often because our perspectives change as we develop, learn about ourselves, and about what different career paths really look like. When I was looking at graduate schools, Mary Firestone asked what I wanted to do when I finished a Ph.D. I started to give her the rote answer about academe and research I'd written for my applications. She cut me off. "Uh, uh. What do want to do when you finish?" Mary saw that I was feeding her pre-prepared bullshit. I responded, "I guess I don't really know," to which she said "That's fine, but you'll have to figure it out sometime." As it turns out, my answer turned out to be true, but it was still just BS at the time; Mary saw it and called me on it (nicely, even kindly, but firmly).

As people grow and gain clarity in their focus, you can help them develop the skills they need to accomplish their goals. Most graduate programs are good at training students how to ask research questions, design studies, and generate results. Programs and advisors vary in how well they train students to interpret results and to develop the stories that grow from them. Programs are yet more haphazard in whether they offer anything approximating training in how to effectively communicate those stories—for example, written, visual, and oral communication. Learning to teach is another key skill, and universities are increasingly offering training for teaching assistants, as well as certificate programs for graduate students (and even postdocs) who plan careers in college teaching.[2]

Beyond that, however, there is a wide swath of activities where universities usually provide limited or no training. Whether we go into academe, government, work with nongovernmental organizations (NGOs), or industry, we will almost certainly have to manage budgets and oversee groups of people. Yet, I've rarely seen training for Ph.D. students in project management, accounting or personnel management—and then only in engineering or

---

[2] For example, UCSB's Certificate in College and University Teaching, which couples coursework in pedagogy with classroom experience. This certifies that someone isn't clueless in the classroom, and that they care about teaching.

professional schools. Some of us figure out how to handle those responsibilities, others of us don't.

Even when such training is available, how many faculty encourage students to take advantage of it? I've known faculty who feel that anything that doesn't directly generate papers is worthless—a waste of time that should be discouraged. I think that's selfish; treating students as the equivalent of "child labor," that is, as tools to be used until they "grow up" and can be cast off to fend for themselves.

8.2 Mentoring is not Altruism, it's Mutualism

The mentor–mentee relationship is always symbiotic, but "symbiosis" means "intimate association of two or more different organisms, whether mutually beneficial or not." It isn't, however, always *mutualistic*,[3] although it should be, and done well, it is. Students choose to work with well-known and accomplished scholars to "ride their coattails," figuring that working with the best will help launch their career. But equally, students push those coattails along! I invest heavily in my students: time, money, emotional energy, and career support. I expect them, in return, to also invest heavily through their creativity, careful thought, and hard work. That produces the work that advances *both* our careers. That may sound very "transactional," but it shouldn't feel that way—my students are my team, and most have become friends.

Done poorly, however, the relationship can become parasitic—and either partner can be the parasite. Faculty become parasites when they use students as workers to produce data and papers, or as teaching assistants, but don't provide the support and career development. Students become parasites when they just ride the coattails of their adviser and others in the group—working hard, perhaps, but without taking the creative lead or making a substantial intellectual contribution.

Don't be a parasite—take care of your people—but don't let students suck you dry either. Dragging a marginal student through their program is not doing them a favor. If this isn't the right path for them, give them the opportunity to realize that. Otherwise, you're likely setting them up for greater failure later when others expect them to function at a level of independence and creativity to which they may not be suited.

---

[3] "The relationship existing between two organisms of different species which contribute mutually to each other's well-being." *Oxford English Dictionary*.

8.3 Advising Approaches

Although I frame one overriding goal for mentorship, there are many ways to achieve that goal. You have to find an approach in synch with your personality, but recognize that what works with one student might fail horribly with another. Regardless, your trainees are *yours*—your responsibility to figure out who they are and what they need to get to that right place.

Achieving that goal isn't about planning, but about trial, error, and experimentation, because every trainee is different and needs different things. A useful concept comes from the military, where plans start by defining the *Commander's Intent*. Despite the chaos of war and the natural intrusions of Murphy's Law, what is the central objective you still must strive to achieve? Then you can identify the more specific targets and approaches—in academe, that means figuring out what that right place is, identifying the skills and knowledge needed to reach that place, and then helping trainees develop the record of accomplishment that will enable them to secure their goal. There are many ways to accomplish those goals, but don't ever forget the key *mentor's intent*: help your people get to the right place for them.

Some people are widely recognized as being wonderful advisors—God's gift to graduate students. Their students are happy, productive, and successful. Equally there are faculty whose reputation is as the Devil's curse on graduate students. They hold students to impossible expectations, are harshly (and often personally) critical, and just generally make a student's life hell. Most of us, of course, lie somewhere between those extremes; we provide a smorgasbord of traits and approaches our trainees might selectively emulate or avoid.

A student's first job is to learn how to ask and answer good questions; that requires space to struggle and explore. Being too directive stifles creativity and growth, but leaving a student floundering leads to wasted time, frustration, and failure. Hitting the balance between those extremes can be a challenge given that it differs between students, and shifts over time for any particular student as they develop. I once described Mary's mentoring style by saying that I felt she would let me take enough rope to hang myself, but if she saw me tying a noose, she'd step in; and she was *always* there if I had questions. I valued that approach and it worked well for me (her approach with other students may have differed). But I know faculty who wouldn't notice that a student was struggling until they found the figuratively dangling body, and others who ride such close herd on their students, they don't let them "play with rope" at all—it's dangerous.

Which specific approach works best comes down to the three terms in the equation: advisor, advisee, and the nature of the interaction between them.

We can't control who we are, but we can work to develop healthy interactions. That doesn't mean that advisor and advisee won't have problems—I figure all students want to throttle their advisor at some point, and vice versa. But in a productive relationship, such feelings are rare, passing, and the exception, not the rule.

We struggle to figure out our people, and we succeed (or fail) with each differently, based on our different relationships. For example, I've known colleagues who could be great advisors for those rare, talented, students who were almost ready to be postdocs when they start a Ph.D., but these same colleagues could be hard on students who needed more development. Some of those students failed, while others were able to move to a different group and did well.

I also have a friend who, as an assistant professor, was frustrated with several students. I reminded my friend that they should realize they'd be lucky to have one or two students in their entire career who were as good as they were themselves—my friend was a rising star in their field and a standout in a top-tier University. Few students are that amazing. I've had several students who are smarter than I am and, who from early on, were clearly destined to be stars. But most students are "just" on a trajectory toward a successful, rewarding, and productive career, possibly as a faculty member at a prestigious university or at an NGO or agency. Expecting a new student to function at the level you expect of a postdoc, or to think and work like you, is unrealistic. Also, unfair. New students take time to develop and often struggle during that period. It's our job to struggle with them to help them grow, deal with the stresses, and reach their potential.

8.4 Mentoring to Enhance Diversity in Academe

Remember—it's a *University*. "The whole of creation." From Chemistry to Classics, Ecology to Engineering, Geology to German, and Physics to Philosophy. The sweep of human knowledge and thought. Given the challenges our global society faces, we need the creative output of all of humanity to solve them. Universities must engage and nurture that talent; we must do a better job of increasing the diversity of our populations.

As I first draft this during summer 2020, society and academe have hit a tipping point. Universities have made progress over the years in increasing the diversity of our student and faculty populations—but we remain far from where we need to be. We cannot fulfill our role in society if we are not representing and serving all of society. Yet, we are dealing with the fruit of a social structure that struggles with problems that are deeply rooted. Communities that have been historically disadvantaged mean fewer students making it to college, fewer with the academic backgrounds to thrive in top-tier universities' fast-paced curricula, and fewer looking on academe as a career path for them.

Students from communities that are less economically established and secure also may face social expectations to get a job and to capitalize on their college education. Take on yet more years of "school"? What will that lead to? Even in the absence of overt discrimination that would kill opportunities, community social pressures can be powerful and pervasive. My father was a bright kid from a working class New Jersey Jewish family, but in his family, success was "My son the Doctor."[4] Embarking on an academic career, with its long development time, modest salaries, and high risks, calls for passion for the field; but passion alone is rarely enough—it needs to be teamed with either a strong social support network (financial and emotional) or the personal courage and determination to break from those social and family norms.

There are smart, talented people everywhere. Only a small fraction of those—in *any* community—have the ability, creativity, and passion to get Ph.D.'s, and a smaller fraction yet to become university professors. When it comes to expanding diversity, academe thus has a pipeline problem. Worse, a number of mechanisms create obstructions in that pipeline—these include everything from overt discrimination, to no-faculty-who-looks-like-me identity issues, to social isolation, to parental and social expectations. These squeeze that limited flow to a trickle. That, in turn, feeds back on, and intensifies, the identity and community issues, adding the "system" to "systemic racism"—even in the absence of active oppression, social dynamics can still suppress the success of students of color and other marginalized groups.

My focus here, however, is not those larger systemic issues of diversity, equity, and inclusion (DEI); but on their more immediate face in the individual one-to-one relationships involved in mentoring. I've defined my philosophy of mentoring as "It's your job to help people get to the place that is right for them." But how do you figure out what that place is, and how do you help people get there? How do you help people who might not initially aspire to things that their talent would allow them to excel at, but which they might well grow into, discovering over time that it *was* the right place?

One key is to have the conversations and to *listen*. But when people's lived experience and social backgrounds differ from our own, it may be easy to listen, but it can still be hard to *hear*—to understand what their words mean. It can also be harder to have the conversation with people who don't share your

---

[4] Anti-Semitism in the 1940s didn't help—Jews could get into medical school, but a Ph.D. program?

background and experience. I have no magic answers. In fact, there are no magic answers. If there were, we'd have waved that wand years ago.

The nonmagical answers involve a mix of things, including to work harder at listening, to be sensitive to different backgrounds and communication styles, and to develop a group and community where diverse voices are welcomed. Equally important is to support students in forming their own support networks. Everyone needs places of sanctuary where they can be at ease and speak without having to filter their thoughts or think about how to express them—or perhaps how to express them safely. I can't imagine the lived experience of our students of color at UCSB; my own experiences can only provide the palest, *external*, glimpse at the issues they face—but nothing about what they feel like internally.

When you interact with a student who has more academic talent than ambition, how do you figure whether their modest ambition is based on past (incorrect) messages they may have received about their real talent level or on the opportunities that they should think about? It's never appropriate to try to force someone into a path that they don't want to take, but how do you distinguish between a true "this isn't for me" and a not-so-true "I've been told that this isn't for me"? If it's the former, it's our obligation to respect that; if it's the latter, we should do what we can to unteach that lesson. Only then can we (both the mentor and the mentee) assess the right place for the trainee.

And, of course, if those students look at the faces on our faculty rosters, they might get the message that this isn't the right place for them—because there are likely still few people on that page who do look like them. Historically, the "right place for them" wasn't the university—university faculty were mostly white and male.[5] That is changing, albeit too slowly. But that means when we take students of color or other minoritized groups into the academy, they may look around and feel like it is not the right place for them.

Part of mentoring our trainees ties into the larger institutional and societal issues—we need to make our universities "right places" for all people who have the ability and desire. But the deep social feedbacks have made diversifying our institutions challenging, in part by creating a Catch-22: When the students don't look like the faculty, how do we create a faculty that looks like the students?

At the level of an academic department and a particular job search, we can only hire people who apply and are qualified. We only have immediate access to the "flowers" of the problem—at that point, we don't have direct access to its roots. We have been working over the years to change that, but we can't just

---

[5] We could add heterosexual and Christian to that.

uproot the forest. Universities change slowly and even when we push, society pushes back. The pipeline problems are deeply buried in societal structures, and even if our intentions were perfect, our ability to create desired outcomes is not. And our ability to do so quickly is even more constrained. We can't uproot the forest, but we can, and must, plant seeds, fertilize them, and protect them from toxins and pathogens. Over time, that will create a more healthy, productive, and diverse community.

When you are working with junior colleagues (from undergraduates through junior faculty), one piece of advice would be to encourage them to find a wider range of mentorship—there are issues trainees or junior colleagues may not want or be able to talk to you about. Encourage them to find the people with whom they can talk. One outstanding resource is the National Center for Faculty Development and Diversity (NCFDD; FacultyDiversity.org). In their materials, NCFDD notes that to expect anyone can get all of what they need from one single guru of a mentor is a fantasy, as I discussed in Chapter 2 (Table. 2.1). We each need to develop our own circle of mentors, each of whom may center on one facet of the overall structure. As advisors, we need to recognize and accept that we need to support and encourage our trainees to develop their own support networks.

8.5 Being a Role Model

Part of being a good mentor is being a good role model, setting an example and demonstrating through your actions and behavior how to be an academic. Yet, I think many professors, even many of the best mentors, can be poor role models. The faculty in graduate programs, people students most want to work and study with, live lives that can be hard for us to identify with when we are starting out.

I once attended a panel at a national conference where several of the most accomplished women in the field discussed how they'd built their careers while juggling family and life. Afterwards, a friend commented, "Yeah, but what about us mortals?" That brought nods from several other women. My friend, who is quite successful herself, still couldn't identify with the superstars on the panel. They weren't credible role models for her. I had a similar experience with my advisor. Part of what drove me to a crisis of confidence was thinking, "I'm never going to be a Mary, so is there any point in doing this?" Mary was still an assistant professor, but was already awesome in a way I could never envisage being. Her career wasn't a plausible model for me.

Equally, I looked at how senior people in the field lived their lives—if this is Tuesday I must be in Washington. I said, "I want to not do that. I want to do science. I don't want to live my life on airplanes and in meetings." Of course, some years later, when I was getting home from yet another trip, I looked at what I was doing and asked, "Uh, Josh, isn't this what you said you wanted to not do?"

This creates a conundrum—the things we do to build our careers, and to support our trainees, also can make us poor role models for them. How do we address that apparent contradiction? First, I think, is to let trainees see the specifics of what we do and why, even though they may not want the whole package. The better they understand us, our decisions, and our lives, the more accurate their mental model may be. When I was a student, I knew I wanted to be a scientist and to do research. I thought I wanted to be a university professor—but I didn't know what that meant. As my understanding of what it means to be an academic has grown, so too, has my willingness and desire to take on those service and leadership roles. I have come to value and enjoy the things I had thought I wanted to not do!

Second, make it clear that our students and postdocs don't have to become *us*. Becoming *us*—a professor at a research university—is not the measure of our trainees' success. Getting that message across may take more active conversation than we might anticipate because it's very easy, looking from the outside, to get the wrong impression of what our "parents," academic or biological, expect.

That was my story with my father, who was a successful psychoanalyst; he spent a lot of time in his office seeing patients or writing. Up through my twenties, my perception of family expectation was for driven accomplishment—but I struggled and never felt I lived up to it. It was only in the last years of my father's life, when my parents visited me after I had become an assistant professor at the University of Alaska, that we had a deeper conversation about his life. He spoke of feeling that he could have reached a higher level of crossover success into the wider public arena, but that he made choices to family and to have the full life he wanted—he accepted the tradeoffs. I'd never realized! I loved my father, but those last conversations transformed my understanding of him, his life, and of course, of my life. I'd struggled with my incorrect perceptions of my parents' expectations, based on my misrepresentation of my father as role model and of what he was really modeling. I had a somewhat similar experience with Mary, but I'm thankful it didn't take me twenty years to figure out I'd been misreading her "mentor model." We know why we make the choices we make in our lives. Our "kids," on the other hand, see things from their perspective, and they can easily draw incorrect or incomplete pictures, and so easily misread the model.

The only way I know to help our students avoid misreading us as role models is to have those conversations earlier and often. I have made choices about my life and career that felt right for me when I made them, and which have taken me to places that I still don't quite believe. How the hell did I get here? Many of us suffer from chronic, and acute, Impostor Syndrome. But our students likely don't know that!

You don't need to join United's Million-Mile club to succeed in academe. You don't need to become your advisor to be successful and content. I am now a professor at UC Santa Barbara, one of the top universities in the world. But I started at the University of Alaska Fairbanks, which is not. Had I stayed at UAF, my career would have taken a different path, but that would have been fine. I was happy there professionally—I left to be with Gwen. Various of my students and postdocs have taken jobs at universities where the pressures are less intense, several have taken jobs at teaching colleges, while others have left academe entirely. They're happy and productive, in the places that are right for them. That is what we should want for our people.

Help your trainees recognize that the advisor they see now wasn't always that person, and may still not feel themselves to be. Growing from intimidated Ph.D. student to Full Professor isn't a magical jump, but a slow evolution. Your trainees' lives will take surprising twists and turns, and likely some of those students will surprise themselves one day by looking around a meeting room and asking, "What the hell am I doing here?"

9
Mentoring 2

Specific Challenges

The stress of grad school can drive anyone temporarily mad
Jonathan Kellerman (2003)

Graduate school is hard. It involves long hours, intellectual challenges, personal transformation, and stress. Helping students manage these, and so to succeed, is the essence of being an advisor. When students fail to complete, it is usually because they get hung up on one of those factors, and the problem then ramifies through the others—they become toxically intertwined. Recognizing the various elements and how to help students sort through them is at the heart of successful mentorship.

9.1 The Routine Challenges of Graduate School

9.1.1 Workload

Being a scholar is ultimately about creativity, not drudgery, but drudgery has its place. After designing the perfect study, you still have to buckle down and *do it*—all successful research involves some nearly insane effort: all-nighters in the lab, brutal weeks processing samples or camped out in the archives, writing marathons. When I worked at the Ecosystems Center, we'd come back from field sampling and I'd be facing fifty or more hours on the gas chromatograph. I would push through in less than four days, and my friends learned to stay away toward the end of that time, when I was, to put it mildly, tired, grumpy, and stir crazy. I loved it. Often, these pushes become a group party with lab members joining in; as professor, my most important job then may be just to make sure everyone gets fed—go out and buy the pizza! It was the same with my advisor, as I'm sure it was with hers. The push to finish writing a dissertation is also often a high pressure slog: a marathon run at a sprinter's pace.

If a student isn't willing to invest the time, the solution is painful but at least clear: get rid of them. You can't do this on a forty-hour work week. Being an academic is about who you are; if a student wants to do a Ph.D. as just a job, they're doomed.[1]

But those killer pushes should be the exception, not the rule. Life shouldn't be defined by the high-intensity academic drudgery. Our careers are framed by our best work, not just its bulk. Ultimately quality is more important than quantity as I noted in *Writing Science*: "Einstein's face isn't on t-shirts and advertisements around the world because he published a lot."

9.1.2 Intellectual Struggles

The intellectual challenges grow from the simple truth that scholarship is about *ignorance* (Firestein 2012).[2] It's about the gaps and errors in our knowledge and how we try to fill and fix them. What's a good question? How do we answer it? These are *hard*. And hard things can be frustrating and stressful. Intellectual challenges are the most straightforward of the mentoring issues we face, because they are the most organic to what we do. The intellectual challenges should also be fun and rewarding—they are what drew us to academe in the first place. The harder issues are the personal struggles that emerge from the long hours and intellectual struggle.

9.1.3 "Normal" Work Stress

Frustration is organic to research. Ideas don't pan out, experiments fail, and Murphy's Law rules—those ion-exchange resin bags I used to measure N in the soil? They smelled intriguing to a passing bear; now the experimental site is a disaster zone strewn with ion exchange pellets and bear scat. Oops! Learning how to manage the resulting stress is central to becoming a professional. As mentors, it is our job to help our students manage the process.

Students have the benefit of focus; they have one project and one goal: *finish it*. But that focus can exacerbate stress—what happens when your project is

---

[1] Masters students are different. A Masters can be more about exploration or job training; it shouldn't require the same level of personal commitment as a Ph.D.; shorter in years but potentially also in the level of submersion.

[2] A brilliant and eloquent little book. I'm jealous I didn't write it.

struggling? When I was a Ph.D. student, if my project wasn't going well, it was depressing, so I'd be less inclined to work. That meant I was less likely to solve the problems, and would have less to show for myself. So, I'd be less inclined to go talk to Mary for support and advice on how to solve my problems, and the project would continue to go poorly, driving the depressive cycle. The opposite would also occur—if I got a cool result, the second person I'd want to tell (after whoever happened to be standing nearest) was Mary. So I'd go talk to her, get positive energy from that conversation, and get even more fired up. It's a positive feedback loop that can run either up or down, leading to cycles that can become somewhat manic-depressive.

The natural stresses of graduate school often grow from uncertainty and crises of confidence—This is hard, can I do it? Our advisors' accomplishments also can pour fuel on those uncertainties. Overcoming that uncertainty, and powering through graduate school anyhow, calls for one of two traits: confidence or courage. I actually prefer working with students who rely on courage—confidence can morph into overconfidence and hard-headed certainty. Courage, on the other hand, is the drive to push ahead even if you are not confident. Over time, accomplishment may seed actual confidence. Different students rely more on one trait or the other. More of my female students have seemed to rely on courage, some almost ferociously so. Or perhaps traditional gender roles make it harder for men to show their lack of confidence, and so they mask insecurity behind false confidence? Regardless, graduate school will likely strain the bounds of both confidence and courage, leading to stress and angst.

To help students through angst-laden crises, aim for balance and presence, and provide a solid base of support, as Mary did for me. Too much pressure might inflict damage, and lead a student to quit. Too little of a mentor's presence might allow stresses to fester, and equally lead a student to quit. A student once asked whether I was a "hands-on" or "hands-off" advisor. I pondered and answered that I try to be a "hands-there" advisor, available and supportive, but not domineering.[3] Try to help students through their challenges, and thus to develop both confidence in their abilities, as well as the courage to push into the unknown and the insecure.

One of the tools to help students manage the stress and succeed is a supportive community—a research group and department that draw out the best, developing a culture of mutual support. But I know of places that beat down all but the "toughest" members, developing cultures that can be toxically

---

[3] You'd have to ask my advisees how they would characterize me—and my interactions have differed.

competitive. You may not be able to single-handedly change a department's culture, but you do shape the culture of the group you lead.

9.1.4 Group Dynamics

Anytime you combine groups of people with work and stress, tensions and conflicts will arise. Most commonly, this will just be an occasional spat: conflicts over resources, impatience with each other, or just venting pressure. I once observed two of my students having a brief squabble and was surprised—I hadn't realized they had problems with each other. I asked, "What was that?" They both acknowledged that it was just a brief one-off. I was relieved, they're adults and can sort out the little stuff. But if little stuff becomes bigger, it can destroy the group and harm all its members. Then you need to become a diplomat, trying to defuse conflict and smooth relations. Once again, I have no magical approaches for doing that—everybody's different and all you can do is hear them out and try to help them sort things out. In the natural sciences, the most common cause of conflict is resources: Who gets to run their samples on the HPLC tonight? Such problems can usually be addressed by having clear approaches for scheduling and prioritizing: Is there a sign-up sheet? Regular meetings where you can discuss who's planning, what can avoid scheduling conflicts and defuse problems. Overall, most conflicts can be managed through regular communication, such as weekly group meetings, or by wandering into the lab to keep tabs on how things are going—if tensions are developing, can you defuse them before they grow into something serious?

9.1.5 Life Balance

One important aspect of developing a productive culture is supporting the idea that both life and scholarship should be fun. A happy group is a productive group. There should be time for having fun and letting off steam.

Life balance involves both how you approach time during the work day and what you do outside of it. There are times when we need to take a break. I once walked into my lab and a student was putting on his wet suit to go surfing. He said, "It seems every time you walk in, I'm heading to the beach; you must think I never do any work." I responded by saying, "Let me tell you a story: When I was a student, it seemed like Mary Firestone would inevitably walk in when we were goofing off, and we said to her, 'you must think we never do any work.' She responded, 'Let me tell you a story...'" We often need a break at around the same time; for the lab group that means goofing off with each other (or in Santa Barbara, going surfing). For me, alone in my office, it means wandering down to the lab to see what's up—"Hi guys, I'm bored, entertain me!" It shouldn't bother you that a student keeps their surfboard in the lab, as long as they produce.

Staying energized and motivated over the years of graduate school requires maintaining life balance. It should not only be OK to have outside activities—it's valuable, perhaps essential. They keep us sane and happy, providing a community with perspectives outside our texts and test tubes. I don't think I'd have survived my Ph.D. without the Berkeley Morris Dance team[4]—Tuesday evening dance practice was a key part of my life. Even during the final weeks of my Ph.D., when I was in a desperate push to finish, often sleeping four hours a night, I at least showed up for ice cream after practice before going back to my computer; my team members were a vital emotional support.

As advisors, we must support student health and well-being—not just data generation. Students who are stressed out are likely less productive and less creative, as well as more likely to make mistakes. They aren't doing well for either us or themselves and are at risk of burning out.

9.1.6 Transformation: Academic Adolescence

Several points in a graduate career are notably stressful. The most obvious is the push to finish: deadlines loom and job prospects may be hazy. But the harder transition is usually at the halfway point and is associated with advancing to candidacy. It may be that qualifying exams are terrifying, but it is rarely the actual exam that is at the root of the terror. Rather, it is what that exam is supposed to demonstrate.

We come into graduate school as students; we go out at the end as professionals. These may be the same physical body, but in important ways are different "people." The transition from student to professional can be both intellectually and emotionally trying. It goes beyond just knowing more—it's about thinking and viewing the world differently. I call this phase "academic adolescence." For many it can be as hard as actual biological adolescence.

Not only is academic adolescence a hard time for students, it can also be a hard time for advisers—just as our own adolescence was hard on our parents. Students are struggling with becoming academic "adults." As advisors, we

---

[4] Morris Dancing is a form of English folk dancing. It's great fun, but hard on the knees.

begin to expect them to function at that new level, but they don't always, still sometimes making errors that we expect them to have grown past. Helping students through this transition can be tricky—in part because we can be frustrated, annoyed, and dismayed, too.

One of my students had a particularly rough transition and was talking about quitting. I, on the other hand, was thinking "You can't quit—I'm going to kill you first!"[5] I was irritated and angry at some choices this student had made, while they were frustrated that things weren't going well and pieces of the project hadn't gelled. I restrained myself from saying something harsh, if only by the skin of my teeth, and managed instead to say "That's an important question, but now is not the time to answer it—we're both tired and stressed, not a mindset to make major life-decisions." As it turned out, we never had to revisit the question. That moment was the crisis; the student rolled past it and after that, everything started clicking, leading to a lovely Ph.D. and a productive career. But it was, as Wellington said of the Battle of Waterloo, "a close run thing." Had I slipped and allowed myself to vent my emotions, this student might have quit, leaving a productive career, and our friendship, stillborn.

My personal crisis wasn't quite so exciting; I found a year-old "to do" list, but couldn't check anything off as completed. I was frustrated and felt like I wasn't accomplishing anything. That was the one time I went to Mary just for reassurance that I was doing OK. She didn't give me a maternal "there, there, you're fine" talk. Instead, she "falsified my hypothesis" by showing me the "data," reminding me of what I *had* accomplished. She gave me the reassurance I needed, but did so in a way that was in line with her nature and our relationship. Thank you, Mary.

9.2 The Non-Routine Challenges of Graduate School

9.2.1 Mental Health Challenges

Some measure of stress is normal in graduate school. Sometimes, though, students go beyond the struggles of dealing with normal stress and suffer from real mental health problems. How do you deal with a student who is suffering from depression, who if you push them too hard or give them bad news, might actually suicide? That might seem extreme—maybe even unrealistic—but it isn't. Depression is common among graduate students—studies suggest that

---

[5] I should note that I meant that in a purely rhetorical sense as a personal, internal, venting of frustration.

one-third or more of graduate students suffer depression. One study found that more than 95 percent of students report "feeling nervous or worrying a lot" (Garcia-Williams et al. 2014). Notably, 7.3 percent of students reported thoughts of suicide and 2.3 percent actually made plans to kill themselves. I had stomach problems that evaporated when I finished my dissertation and started a postdoc—but I was moving from Berkeley to Scotland, so you can't blame the food. With so many students suffering mental health problems, you will experience suffering students during a career as a teacher and advisor. Accept it. Expect it. And, to the extent you can, as Garcia-Williams and colleagues suggest, prepare for it:

> Educators should be trained to recognize the factors associated with suicidal behavior among graduate and professional school students.
> Educators must be prepared to link graduate and professional school students they are concerned about to appropriate mental health resources.

Emotional crises also can be catalyzed by external forces, such as a close family member being a victim of violence. How do you support a student in such circumstances? Back off and give them space to heal? But how much is the right amount? How do you remain personally and professionally supportive? How to know when to start re-engaging? How much, and when, to push to test the waters—is it time to nudge them forward?

To help a student in crisis, it would help to have training in counseling or psychology. But I have none! I had never realized that being a professor comes with a side job as a mental health amateur. Yet, we are cast in this role, and we should be bound by the Hippocratic Oath: First, do no harm.[6] But how do you avoid doing harm when you don't know what you're doing? I'll suggest two steps: Step #1: Fret, worry, and lose sleep yourself. Ask yourself: "How can I support this student? How can I help them find a safe path?" Step #2: Get help for both yourself and your student. We aren't alone in these situations: as faculty, we are part of a large university team. Talk to your departmental graduate advisor, who can offer advice on navigating rules on timelines and finances. The campus graduate division and student health program have people with training and have dealt with all sorts of student crises—*use them*. They can advise you, and they can offer services to the struggling student.

---

[6] The language in the original (translated from Greek) is actually, "I will abstain from all intentional wrong-doing and harm."

9.2.2 Emotional Entanglements

Predatory professors viewing the coeds as a personal playground is a classic trope, and although such people still exist, campuses have gotten better about weeding them out. Good riddance. But there is an equally old trope of professors and trainees forming romantic partnerships. I have several friends who married their advisees or advisors (and those relationships are often stable and happy). Our trainees are smart, creative, and passionate about the same things we are; we spend a lot of time working closely together. Hence, it is unsurprising to find yourself attracted, and equally unsurprising that romantic attachments sometimes develop.

But even when such relationships are happy and healthy, they are problematic—it would be hard to deny favoritism if you are sleeping with one of your research team. If a relationship sours, however, the dangers become severe. When I attended my first training session about sexual harassment at UCSB, the woman who ran it, Paula Rudolph, said that the hardest cases she had to deal with were between assistant professors and their postdocs, unmarried people close in age and interest. But because of the employer/employee relationship, in the midst of a breakup something as normally human as simply saying, "I love you—don't leave me" can cost a professor their job and career—such words can constitute legally actionable sexual harassment.

So while having feelings for your trainees may be unsurprising, tamp them down. Your responsibility is to help your people develop their lives and careers. Unless a relationship with you is the thing that achieves that goal for them, allowing a relationship to develop violates mentoring and professional standards. If a relationship appears to be fermenting anyhow, talk to your Title IX coordinator,[7] *immediately*, to discuss approaches to manage the situation; there may be ways to shift the supervisory relationship to limit liabilities.

9.2.2 Students Who Aren't Going to Make It

Although almost all of us struggle at some point as students, most of us still break through and complete degrees. But, some don't. Students may fail to complete their degree either because they come to realize that this is the

---

[7] This refers to the U.S. law that deals with sexual discrimination. Outside the United States, talk to your department chair or other appropriate official who can help restructure the working relationship. You should not be responsible for signing off on your romantic partner's Ph.D.!

wrong path for them intellectually, or because the emotional stresses of graduate school overwhelm them.

Being an academic is not just a job; it's a calling, probably not unlike being called to the priesthood. It's not just what you do, it's who you are. Thus, graduate school is not for everyone—not even for everyone who loves scholarship. The fundamental premise of this book is that being a scholar is distinct, and more, than just doing scholarship, whether that scholarship is history or histology. As a student discovers that they don't want to be an academic, a crisis is sure to follow. Some students go into graduate school without knowing what it is really all about. It is easy to love knowledge; the "science nerds" we knew (or were) in high school loved science, but how many had ever done science? We were learning about science and scientific knowledge, but we weren't doing science and we weren't being scientists. That can remain true through college: classes teach students what we already know. But science isn't about what we know; it's about what we don't know (yet). Science is about the Terra Incognita beyond the borders of knowledge. Science is an approach to create knowledge, taming the wilds of ignorance and putting it "under the plow," opening up new fertile and productive territory.

Some people, even those who love the beauty of knowledge, are not suited to being scholars—they like the *knowing* rather than the *finding out*. Yet, scholarship is a struggle not just to answer questions, but first to figure out the right question to ask! Some students, even very smart ones, don't have the right type of creativity to succeed in research.

Often it takes months or years for a student to realize that a Ph.D. is the wrong path. After years of loving their studies in college, envisioning a career, and then applying to and beginning graduate school, students start with passion and enthusiasm. The realization that you are in the wrong place comes slowly and with frustration, stress, and emotional trauma.[8]

I went through that in college and it was not fun. I had known, even in high school, that I wanted to be a chemist. In college, I majored in chemistry and didn't emphasize anything else—no biology, no environmental science, nada. Senior year, I was working in an organic chemistry lab, and I came to realize that although I loved the mechanistic perspective of chemistry, I didn't feel emotionally connected to the questions on which I was working.[9] I couldn't see going forward on that path. I was depressed and didn't do very well.

---

[8] Relationship-breakup or divorce-level pain.
[9] Or maybe the mix of glacial acetic acid, ether, and mercaptoethanol made me so sick it put me off the entire field.

I graduated with no idea for a future. I could see the closed door, but couldn't see how to open it or what might be on the other side.[10]

I didn't have anyone to help me through that realization, making it harder. But your graduate students are yours. For those students who discover they are in the wrong place, it's your responsibility to help them through the process, while making it clear that it isn't a failure, but a recognition. Not having the "calling" doesn't make you a bad person.

Dealing with students who are struggling is emotionally trying; frustration is guaranteed. We take people on with good intentions and hope; when students don't complete, we see that hope go unfulfilled and projects languish. The frustration can be partially mitigated, however, and the misery limited. A student's decision to leave a Ph.D. will almost certainly go more smoothly if you reduce the activation energy for their decision—be supportive of the idea that if this isn't what they should be doing, then deciding to do something else is the right choice, one you support. If your message is that you expect them to tough it out, it makes it harder for them to quit; a student may feel obligated to "live up to their commitments." But, the commitment they should live up to is the one to themselves to do what is right for them.[11] Throwing good money after bad is equally bad tactics in poker and academe. A student who is staying on out of obligation instead of passion isn't likely to ultimately be productive or successful. They are likely also draining your time and energy. If you really need someone to wrap up a project, work out a fair employment arrangement instead of maintaining the student fiction.

Sometimes when students feel caught in a stress cycle, their inclination is to take time off and to move away from the source of the stress to re-energize. This is often a bad idea. The support network that can best help a student through academic struggles is their departmental peer group and faculty mentors.[12] Family and hometown friends love you, but they aren't likely to provide the academic energy. This may be particularly true for first-generation university students, whose family have little experience with what it means to be a graduate student. In my experience, students who take time off and go home to recharge rarely come back. Rather, we need to provide the support, community,

---

[10] I was lucky. My brother David was working at the Ecosystems Center in Woods Hole, Massachusetts. He helped set me up with Jerry Melillo, who needed a lab tech. I discovered how ecosystem science used a chemist's mechanistic approach to answer questions that appealed to me, so I applied to do a Ph.D. in soil microbiology.

[11] A student should make a commitment to their advisor, but to work hard and creatively, not to waste their advisor's time and energy by powering through a program that is wrong for them.

[12] Albeit, a student's advisor is often the perceived focal point of the stress. You can encourage a student to connect with other mentors, particularly those who aren't going to have to sign the dissertation.

and resources within the academic system—we may not experience the stress internally the way this student does, but we have faced parallel struggles.

Perhaps the most frustrating students-who-aren't-going-to-make-it are those where it's pretty obvious from early on that there is just something missing—the passion, the energy, the ability to operate in that space beyond the edge of knowledge, excessive perfectionism, or just the ability to dive in and do it. How do you handle such a student? I've known colleagues who would just cut them out quickly, arguing that they weren't just cutting their own losses, they were also cutting the student's—minimizing the time and energy the student was "wasting." I understand that approach, but "firing" a student can inflict psychological and emotional damage. I see that as incompatible with the philosophy of helping people get to the place that is right for them. Having accepted a student, they are ours. If it's clear to us that a graduate program is not the right place for them, we still have an obligation to work with them until they come to that recognition on their own. Or, if a student really won't see the writing on the wall, that is what qualifying exams and committees are for. If a student fails their exam, it's because three or more faculty all agreed that the student isn't where they ought to be. That helps defuse a potential individual conflict. Also—you might have been wrong in your initial assessment! The student might need a bit more development time before the pieces click. Sometimes that might mean finding a different advisor: I had a student who did a fine Ph.D. with me, but they switched to my group because they weren't connecting well with a colleague of mine. My colleague could be a good mentor and had succeeded with other students, but this relationship just didn't work. Whatever it takes, it's our job to support our people while they find their path to the right place—even when you might be frustrated and annoyed and even if that means some other research group.

9.3 Postdocs

The challenge with mentoring postdocs is grounded in their definition—a postdoc is defined by what they are *not*. Postdocs fall into a variety of specific titles, but they are all temporary; a postdoc has only a few years to gain experience and develop their professional credentials. A postdoc's job is to get a job.

Mentoring postdocs is in many ways simpler and more straightforward than mentoring students—they've survived academic adolescence and proven that they have the skills to do research. A postdoc is ready to step into a senior role in a research group. The greater challenge in mentoring postdocs is that their focus tightens: they need to get a job and your job is to help them do that. But as soon as they get an offer, they may be gone. Hence, designing a good postdoc project is different than designing a good doctoral project.

In my experience, postdocs who leave to take nonacademic positions almost never get the papers written. And postdocs who take academic positions get distracted in their first years—they may need the publications but struggle to find the time to write them. If a postdoc gets an academic job, though, they can often defer starting that new job for some time—offers are usually made in the Spring for jobs that start the following Fall, but which can often be delayed further to wrap up the postdoc. When postdocs are looking for nonacademic positions, though, their departure may be abrupt. Industries and agencies usually recruit because they need someone now—or at least as soon as possible. That means a postdoc might well uproot and move with only a few weeks or months lead time—short enough to leave projects incomplete, data unworked up, and papers unwritten. Congratulations! . . . but, crap.

There is no simple solution to these challenges. Mostly it comes down to discussing the postdoc's career visions early, so that you aren't surprised and can design their project to reduce its vulnerability to the postdoc's plans. Also, as they prepare to leave, discuss in detail what you expect them to accomplish to wrap up their current project—but also remember what I have to say about "plans" in several places in this book—no plan survives. Drawing back a former postdoc to focus on their postdoc project can be difficult, and it only gets harder as time passes.

Overall, the solution to all these specific mentoring challenges is to be "hands there"—available and open—and to do what you can to help them get to the place that is right for them. Recognize that when trainees are struggling, we are not alone and our universities have people who have professional expertise in counseling. Don't be afraid to use them. There's no guarantee that we will always succeed, but we should always try.

10
Teaching

Being Good while Surviving

Those who can do, also teach

This opening quote I borrowed from the preface to *Writing Science*. Teaching is organic to scholarship. After all, your field doesn't advance because of what you have actually done, but by what your audience understands you to have done—what you teach them.

Even if your career leads to industry, government, or a nonprofit, educating audiences will be fundamental. But for those of us in academe, classroom teaching is likely at the heart of our jobs. That is true even in research universities, where tenure and promotion are driven more strongly by scholarship. The job is research, service, and teaching, so ignore classroom teaching at your peril—strong researchers can be held up by weak teaching.

Becoming at least a competent teacher is essential, even though that may seem at odds with the classic stereotype of "great researcher, terrible teacher." People often assume that these abilities are inversely related, yet to succeed in research, you must be passionate about your subject and you must communicate effectively. Those are core skills in the classroom, too. Thus, I see research and teaching skills as positively correlated, albeit weakly. I know many faculty who are terrific at both: creative, successful scholars who deliver passion and energy in the classroom.

What is undeniable, however, is that time is finite. Regardless of whether we are in a research university or an undergraduate college, time is our most limiting resource, and we all have multiple demands on that time—students, scholarship, and service. Each takes time, so trade-offs are real and inevitable. Developing a balance that allows you to succeed in both teaching and scholarship is fundamental to surviving in academe. This may be trickier in a research university where you must excel in research. In such positions, a common hazard is neglecting scholarship to focus on teaching. Teaching creates a daily flow of work, and interacting with talented and motivated students gives strong, immediate, positive reinforcement. It's easy to get caught up in teaching's "has to be done today" tasks and to lose sight of the "it doesn't have to be done today, but my world will end if it doesn't happen soon" tasks that are critical in building a research program. In the uncertain and frustrating world of research, teaching can become like a drug, providing short-term rewards that ultimately are self-destructive. This is what Eric Hayot (2014) describes as "Virtuous Procrastination," a bad habit to which academics can be prone:

> The most common form of virtuous procrastination for well-meaning academics is teaching: I can't write because I have to prep for class; I didn't work today, because I had so many papers to grade.

He adds other thoughts on the subject:

> These things will matter, but they will cease to matter if you lose your job for not publishing enough.
> I am warning you against making teaching a *substitute* for writing.

It is not that Hayot is arguing we should neglect teaching—quite the opposite:

> The idea that writing is real work and teaching is not makes me totally crazy. The students are our work too.

The issue, rather, is time balance, and not using the short-term rewards of teaching as an opiate. To survive and succeed, you need to teach well and effectively, while still developing the research program that will advance your scholarly career. You must manage the time commitments to optimize the balance between teaching and research.

The first piece in creating that balance, ideally, is for your institution to recognize the role of research and provide a reasonable "teaching load," although I despise that term. If teaching is a "load"—an unpleasant burden—quit. A second piece in maintaining balance is a productive research group, which can allow you to focus on teaching and service for periods of time without your scholarship seizing up—this is common in the natural sciences but less so in fields more dominated by an individual scholarship model, such as in the humanities.

For me, the heart of how I develop a teaching/research balance is a simple philosophy about teaching. This distills down to the recognition that as a teacher, you only do two things in the classroom: #1: you motivate learning, and #2: you facilitate learning. Learning is the important thing, and that is what students do.

Of the two, motivating and facilitating, I think motivating is ultimately the more important. If a student really wants to learn, nothing you can do will stop them. Equally, if a student doesn't want to learn, nothing you can do can force them. Realistically, what we can hope to do is inspire students to a higher level of engagement, and show how the material is useful and interesting. We should be able to do that even for students who may not initially be excited by the material (e.g., students who are taking a class because their major requires it[1]); we may be able to engender some real interest among students who are initially uncertain; and for some rare few, we might fire them to enthusiasm and passion for the topic.

Regardless of students' levels of enthusiasm, however, we must help every one of them learn, within the constraints of their abilities and interest. Achieving that is about facilitating learning. And facilitating learning is about building an effective human system within your classroom—one that connects you as the instructor with your students.

Rather than discussing "facilitating learning," we usually instead discuss teaching tools and pedagogical techniques. There is a lot of interest in active learning, flipped classrooms, and so on, which all come down to approaches to facilitate learning. These can help once students are ready to learn, and we should explore such techniques to improve our classes. But technique alone does little to motivate students—that comes from you, the human instructor in the classroom.

Becoming a successful teacher while still succeeding as a scholar is about figuring out which approaches work for you in the classroom. How do you optimize your performance as a teacher, without it leaving you either no time for scholarship, which leads to professional disaster, or no time for life outside of academe—such as hobbies or time with friends and family—which leads to personal disaster. I see three components to being a successful teacher: passion, communication, and pedagogy. By "pedagogy," I mean how you structure the class and integrate specific tools and approaches.[2] I believe that is their real order of importance, although they are interwoven.

---

[1] I took geomorphology as a Ph.D. student because it was required for soil science. I went to the first class grumpy and resentful—I didn't want to take any more damned classes—but Dr. Bill Dietrich made it one of the best, most valuable, and most fun, classes I ever took.

[2] That is not an accurate definition of pedagogy, whose real definition is: "The art, occupation, or practice of teaching." Everything in this chapter is about pedagogy in its broadest meaning.

10.1 Passion

Because I see motivating learning as the most critical thing we do as teachers, I see passion as the most important element in the classroom. Show that you care about the material—how it is fascinating and important—and that you care about the students personally. That will engage them to want to learn.

The most compelling teacher I ever had was in high school. Dr. Szell was a Hungarian emigree who taught chemistry. Dr. Szell wasn't flashy, he didn't emote a lot (though he had a wonderful quiet humor), he wasn't easy (his "Above Mickey Mouse Level" problems could cause nightmares), and he didn't use tricks. In fact, his teaching was pretty straight-up old-school. What he did was to radiate a deep love for chemistry, for teaching, and for his students. Every student in his classroom knew he was dedicated to us. Dr. Szell loved chemistry, and he loved us. We knew it, and we repaid in kind—we fell in love with chemistry and we learned a lot of it.[3] He motivated more students to go on in science than any other teacher I have ever known.

Motivation, even without clear communication or sophisticated pedagogy, can sometimes achieve near miracles. For example, I know someone who gave an almost disastrous job seminar: awesome science but scattered and disorganized. This person was hired anyhow, but there were worries that students might say that proved the department didn't care about teaching. But this professor gets strong evaluations and is adored by many students. How can that be? Well, particularly in small classes where there is a lot of personal interaction, students can't help but be swept up by this person's passion for, and deep knowledge of, the material. Students note struggles with organization and sometimes even clarity, but they forgive that because the professor is great at motivating them to learn—and so they do.

What students won't forgive, ever, is a failure to motivate. Students won't care if you are meticulously organized and use all the best tools, if simultaneously, you give them the sense that you don't care about your material, or worse, about them. In contrast to the previous example, I knew someone who had a terrible reputation as a teacher. This professor gave lectures that were clear, informative, and sometimes even witty; they cared about the material. Yet, this professor's teaching evaluations could be savage. The attitude this professor projected led students to feel he didn't care about teaching, about whether students learned, or even whether they passed the class. I believe the students misread this professor's real attitudes, but rightly or wrongly,

---

[3] It took me four years as a college chemistry major to quench the fire Dr. Szell lit. I still love chemistry and rely on it as a core tool in my research, but I didn't love it enough to succeed as a chemist.

perception matters. Many students thought he didn't care, so they thought he was terrible, and as a result they disengaged and didn't learn well. Ergo, this professor was, perhaps, not "terrible," but he was unsuccessful.

Students have to see that you care. But how do you show that? You can't just announce "I care"! That works about as well as muttering "I love you" to your romantic partner while you're busy checking messages on your phone! Presumably, you care about your material—how else did you get where you are?[4] Let students see that. If you seem bored, it's almost certain that students will be, too. If you actually don't care, why are you still here?

Let your students see you as a person, and not just as a faceless lecturing automaton. There is a reason we don't just record lectures and then hit "play" when class starts.[5] A good movie may be better than a bad play, but there is nothing like the magic of a good live performance. They are fundamentally different experiences, and live performance draws a deeper level of engagement.

The occasional story about your own experiences might seem like time away from describing the biochemical pathway of the day, but those stories show your humanity and connection to the material—and they might stick with students long after. I haven't forgotten Dr. Szell telling us about when he was a student in Hungary and someone accidentally drank the lab methanol. Oops. They had to pour ethanol down him to save his life—create competition at the active site of ethanol dehydrogenase to limit the accumulation of toxic formaldehyde. Little stories like that not only teach something about chemistry but also offer a window into the teacher as a person. Such "Instructor Talk" can build a bond that energizes learning (Seidel et al. 2015).

Students will occasionally strain those bonds of patience and courtesy. We've all heard questions that create an almost irresistible temptation to respond with, "Well, if you'd been here last class, done the reading, or even been paying attention rather than checking e-mail for the last five minutes you'd know the answer!" If your valid exasperation comes across as sarcasm and disparagement, though, not only will that student never ask another question, but other students will think twice or thrice before they ask—even if they have a good question—and they may think worse of you. There are other ways to approach such situations, such as putting it to the class: "Can anyone answer Josh's question?" Most likely someone will. But also consider that maybe you weren't as clear as you thought, and maybe Josh's question wasn't as boneheaded as it seemed?

---

[4] Even if it's the basic introductory stuff in your freshman class. This was novel and exciting to you once.
[5] I wrote that before COVID-19. In 2020, we were pre-recording lectures, but not because we wanted to.

Often, students come up after class with questions, so leave a few minutes to deal with those. Sometimes though, we do need to rush off. When that is true, a few seconds can make the difference between students thinking you're blowing them off because you don't care, and understanding that you really are busy: "I have to be at a faculty meeting in five minutes, so I really do have to rush—can we get together later today or during office hours?" Remember that students are vulnerable to thinking that we see teaching as a diversion from research.

Care about your students and want them to succeed, though we know not all will. Caring, though, doesn't mean being easy or generous in grading. I have colleagues who think that to get good evaluations all you have to do is give a lot of A's, and that evaluations just correlate with the grades students expect. I don't think that's true; at least it's an imperfect correlation. Students like to succeed, and they like to do well, but the teachers we revere are those we learned from and were motivated by, not just those from whom we got an "A." Easy A's are forgettable; great teachers are not. The best teaching evaluations are those that say, "Great class—I learned a ton, but the teacher is tough!"

Then, put on your performance face—you became an academic because you love your subject, so show that passion. That doesn't require hype. It doesn't require play acting (but don't rule that out if it helps). Mostly, it requires letting your true interest shine through. If you have a hard time with this, and I know many people who have trouble performing in public, get help. Your campus instructional development center can help you, but alternatively, find other ways to become more comfortable on stage. Before my knees failed me, I had been a serious folk dancer—in fact, I met my wife on the Berkeley Morris Dance team when I was a Ph.D. student. Morris in an ancient English tradition that involved getting kitted out with bells on my legs, ribbons on my arms, and performing in the streets. Years of making a spectacle of myself in public made the classroom stage less intimidating. Whatever it takes, find ways to learn to engage in public as best you can.

A last element that can create a human bridge is humor. I figure anytime I can get a laugh out of the audience, I win. I don't care if they are laughing with me, or at me. If students are laughing, they are awake, connected, and paying attention. Be careful with humor, though, because a bad joke can fall flat, an off-topic joke can distract, and an off-color joke can offend. So only try humor if you're comfortable with it. And although self-deprecating humor can make you more human and less intimidating, never make a student the butt of a joke. It's easy, but evil, to slide offensive messages under the camouflage of a joke. Spontaneous wit and carefully planned humor can both work, but ideally they should enhance the message.

My most effective spur-of-the-moment humor was when I was teaching general microbiology. I was discussing how bacteria move in response to a resource gradient. I had slides illustrating the run-and-twiddle, random walk of bacterial motility. But I had a big stage and just started demonstrating—I meandered around the stage, occasionally stopping and spinning before setting off in a different direction. That got laughs then, but I also heard about it from colleagues later—the demonstration was clearly notable and hopefully memorable.

In contrast to that was a seminar on Arctic lake ecology by Dr. Mark Oswood when I was still at the University of Alaska. Mark studied the ecology of shallow Arctic lakes that dot the tundra. He spoke about biological productivity in Arctic lakes and the challenges of studying them. You can put on rubber boots and slog out into the water, which allows you to take some types of samples, but at the cost of disturbing sediments. Alternatively, you can paddle out in an inflatable boat to protect the sediments but at the price of sampling from an unstable platform. After discussing these options, Mark closed by noting: "This highlights the ongoing challenge of Row vs. Wade." Thirty years later, I can still hear the groan that echoed through the room—but I also remember the science underlying the joke!

10.2 Communication Approaches

The second element of good teaching is communication. There are a range of skills involved in communicating effectively—you need to have a good story to tell, and then you must tell it well.

The first component of communication is always about story. What is the story you are trying to tell? I discussed the concepts of "story" and of using various story structures to engage different audiences in *Writing Science*, but the core issues are the same in presenting a class: Who is your target audience? What do they know? How do you frame the story? And what goes where to make the story compelling? As you develop a new course, there are actually several stories you need to develop for students. There is the obvious story about the material you will be teaching—the course content. But there is also the story about the class itself and your expectations. What do you expect from students? We define these—the material we expect them to learn and the activities we expect them to do that they will be graded on—in the syllabus and in the first class period, and in doing so, establish a contract between the teacher and the students. If students fail in their part of that contract (i.e., don't do the required work or do it poorly) we assign them a low grade. If we fail in our part of the contract, students will assign us equally poor "grades" through their course evaluations and the word that will spread through the rumor mill.

Then there is how well we tell the story. That is mostly about the "performance skills" I noted above: engaging with your audience, speaking clearly and with life and animation, having good timing (both within each class and across the term[6]), preparing clear materials, and so on. I already discussed some of these skills in Section 10.1 about passion. And I'll reiterate my advice to work with your institution's instructional development program. Even the best authors need editors, and the same is true for teaching—we all benefit from expert commentary and advice. Our institutions probably all have teams who are expert in instruction and can provide that advice. They can look over your materials, come to your class, record a class session, and help you see how well what you are doing is working; they can help you make it work better.

10.3 Content

Teaching is like stretching a rubber band—while you are actively pulling, the rubber band will stretch, but the instant you let go, it springs back. The longer and harder you stretch, the more likely the rubber band will be permanently deformed and won't spring all the way back.

This "rubber band" model fits with the flow of learning in Figure 10.1. We learn by starting with raw data that we convert into information (facts, figures), which we distill into knowledge, and ultimately synthesize into understanding, stretching the rubber band from left to right.

In this model, learning starts at the left, with data and information. It's easy to learn facts—they are mere snippets: the structure of phenanthrene, the date of the Norman Conquest of England, how to conjugate "être." But, the second a student hands in their final exam, they let go of the rubber band and start forgetting—but they also start forgetting from the left.

Facts are easy to learn but even easier to forget.[7] Is phenanthrene two or three rings; and if three, are they bent or in line? When do you use the plus-que-parfait instead of the imparfait? But when you aggregate and synthesize facts, you find patterns and develop knowledge; if you work with that

---

[6] Students so love it when a teacher realizes they're only half way through the syllabus with two weeks left to go!

[7] Specialists in teaching and learning use the term "declarative knowledge" to refer to facts you can recite; this contrasts with "functional knowledge," which you can apply in other circumstances than those in which you learned it.

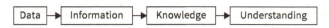

Figure 10.1 The flow from data to understanding. From *Writing Science* (Schimel 2012).

knowledge and struggle with it long enough, you ultimately build understanding. Knowledge and understanding fade slowly; the highest form of understanding is the kind that cannot be unlearned—concepts that change the way we view the world, the so-called "Threshold Concepts" (Meyer and Land 2006). Accept that students let the rubber band go, and design your classes to recognize it.

An effective class is designed around the knowledge, concepts, and understanding you want students to develop. We may present a lot of information, but the information isn't there for its own sake but because it helps students stretch their rubber band. We choose examples and case studies because they illustrate concepts. Evolution classes discuss Darwin's finches and *The Voyage of the Beagle* partially because of the historical importance, but also because the Galapagos offered such a powerful illustration that they led Darwin to develop the idea of natural selection, a threshold concept for sure. When I took biochemistry, we learned that when bacteria are given glucose and lactose together, they don't turn on lactose metabolism until they have consumed the glucose (so called "diauxic growth"). But even the professor didn't really care that E. coli prefers glucose over lactose. No, the value of that example was to show how cells regulate metabolism to maximize their productivity by avoiding making unneeded enzymes. The specifics are trivia—it's the mechanism that matters. Every area of scholarship and teaching faces these issues: What specifics do we work through to ingrain the ideas?

Evaluation material (exams, problems sets, papers, etc.) should emphasize knowledge and understanding over raw facts. I always ask some questions that are more "informational" to ensure that students don't skip over that material in their studying, and that they get into the weeds and work with the details, even though I know they will mostly forget them later. But if we just try to tell students the knowledge and ideas we want them to know, without pushing them to work through the full path to how we get there, those ideas become just facts to be recited rather than concepts to use. To understand natural selection, students need to work through examples, such as those that led Darwin to develop the idea, and they should be able to show that they can use the concepts to explain real world situations.

For students to hang onto material—to truly learn it—they need a framework to tie it together, to provide context and to make the material relevant to their lives, like Dr. Szell's tale of the poisoned Ph.D. student. Any mere collection of facts will bore many of us and will be quickly forgotten. Rather, frame a story that places those facts in context and shows students why they matter. There is no material so fundamentally boring that a good teacher can't make it engaging, yet equally, no material so fundamentally interesting a poor teacher can't make it dull. We all know those people—on both sides. The difference is often between laying out a bunch of facts to memorize, and telling a story. Framing material as story may mean you cover a little less but students will likely learn and remember more—you'll stretch the rubber band further. Give us the hooks to connect specifics to our experiences and interests to make it "sticky."

Humans tell stories. Of all the ways people have tried to distinguish humans from nonhuman animals, storytelling may be the one unique thing. Telling stories is how people teach; that was true back in ancient Greece with Aesop's fables, and it was true for the native Alaskans I had in my classes at the University of Alaska—their Elders told them stories about past experiences to teach kids how to survive in the unforgiving environment of the Arctic.

Framing material as a "story" is at the heart of my book *Writing Science*, and what is true for communicating sophisticated information to engaged experts is even more true for teaching core material to undergraduates. We introduce situations and problems our audience care about, draw them in on a specific issue, show them how it plays out, and then wrap up to synthesize to offer the "moral of the story." This is the opening/challenge/action/resolution (OCAR) model I discussed in *Writing Science*. By presenting information as more than a collection of facts, we give it context and meaning. Raw facts don't stick—if you don't show students what the facts mean, they have to figure it out for themselves; but that's not *teaching*.

10.4 Expectations

My job as a teacher is to hold the bar high and then do what I reasonably can to help students clear it. First, that means showing them where the bar is! Define expectations and make them clear. Then, help students achieve those expectations. The key place where we establish our expectations for students is in the course syllabus. This is a professor's contract with the students, defining what we will cover and expect, and when things will happen and be due. But the "contract" aspect of a syllabus can be a problem. Dr. Helene Gardner expressed it this way:

> I have seen syllabuses with every topic, pages of reading, point-earning event, and instructions in exacting detail. I am so impressed with them. Really. They are

awe-inspiring. Whenever I have published something like them, I have gotten into trouble. In my experience, classes are dynamic, living beings. They just don't eat, drink, and poop on schedule, so, when I have published that the paper on a creative solution to the Lake Nyos disaster is due on Tuesday, the 11th, inevitably the Lake Nyos class sloshes into the next class period, so students aren't ready to start when I thought they would be, so that due date has to be changed to Thursday, the 13th. Everyone will not get the message.

My experiences have been similar: No plan survives contact with the enemy! Or to quote from *Pirates of the Caribbean*, "The Code is more what you call guidelines, than actual rules." Certain dates will wiggle, although key things such as midterm exam dates should not. If you haven't gotten quite as far as you thought you would, you can just test on the material you have covered. But learning objectives define your Commander's Intent, the goals you must strive to achieve, even when the class stops "pooping on schedule" and you need to make some mid-course changes.

Learning objectives should frame the whole course, but also each class period. They should emphasize the knowledge to assimilate, not just the information to memorize. For example, don't just say "Learn glycolysis and the tricarboxylic acid cycle (TCA)," because that mostly translates into "memorize the steps in the pathways," which can be devoid of content or meaning (and are quickly forgotten). A better objective would be "Learn how carbon is converted from plant sugars into bacterial biomass (via glycolysis and the TCA pathways)." Students will likely remember why they are important, and maybe the skeleton of the pathways, even when the details are gone.

Within the context of storytelling, such learning objectives define the overall direction of the story, and they do it right up front. That sets up the class as a front-loaded story, where the point comes early to give the audience a roadmap of the material to come. Hopefully there will be a final wrap-up as well: a strong resolution to highlight the key messages you expect students to take away from the class, and that they can expect you will test them on later.

When we give exams, students expect to be able to see how the questions relate to the material we have covered in class, and particularly how they match to what they had perceived as our key points. They will be annoyed to see a major question about an off-hand topic you spent five minutes on in class. Tests should map onto the learning objectives you've defined. In doing this, many of us give exams that test students' ability to apply material to new problems, problems we hadn't specifically worked through in class. That is appropriate. But if students have never seen that kind of problem, they will feel blindsided. A few, the very best perhaps, may be able to adapt and figure out the links in the short time available for an in-class exam. But most students will struggle and will probably argue in their course evaluations that the tests didn't relate to the class. They may be wrong, but if they can't see the connection, it didn't exist for them. And they'll blame you.

To help students figure out where the bar is and to see the type of logic and approach I am looking for, I post the previous year's test as a practice test and I separately post the grading key.[8] Students may be challenged by my exams, but they aren't surprised by them. I try to highlight core concepts—for example, in my class "Ecosystem Processes," I always test their understanding of Liebig's Law of the Minimum, ecosystem water balances, and carbon allocation in plants, among other things. The questions hitting each topic vary from year to year, but the answers are always grounded in the same core material and principles. They also emphasize the things that I hope students will remember after the class is over (the stretched rubber band). I often use questions that tie the material about ecosystem processes to situations they may actually experience. For example, one of my favorite exam questions is:

> A few years ago, I planted potatoes in my garden. The soil is a heavy clay. To improve the ground, I dug out the clay and replaced it with good topsoil, garden compost, and lots of manure. I also added blood and bone meals for extra nitrogen and phosphorus.
>
> The potato plants came up large, green, and healthy. When I harvested the potatoes, however, the crop was disappointing. There were few potatoes and those that were there were small. There was no evidence of pathogens. Why did this likely happen?

Yes, this little story is true, and my interpretation was that with so much nitrogen in the soil, plants allocated resources aboveground to grow leaves; the plants need to be a little N-stressed to allocate carbon belowground to produce roots and the starch for potatoes. That at least was my theory, but I accept other answers as long as they are plausible and grounded in the processes we'd discussed. In other years, questions about how resources move through plants look completely different, but my bottom line is that I want to see them applying knowledge rather than just regurgitating facts.[9]

---

[8] Practice tests show the kind of thinking I expect; working through them becomes a learning experience on its own; and students feel that by giving them a study-aid, I care about their success. Win-win-win.

[9] The challenge is grading such questions. Answers can be hard to decipher—both intellectually (what do you mean by that?) and physically (written in class in a hurry). Balance how many you ask against the number of students and the amount of grading support you have (teaching assistants or graders). In an introductory class with 500 students, a computer will almost certainly do most of the grading.

Students don't always do well in my class, but I don't get many complaints about being unfair. The one class where I did get criticism was during the COVID-19 pandemic when I went online with pre-recorded lectures. I assigned a weekly low-stakes quiz—two or three quick multiple-choice questions to encourage students to stay on top of the material. Multiple choice questions are hard to write and some of mine were imperfect. Students complained about getting answers wrong because the question was unclear. The questions they missed because I wrote them poorly had no influence on their overall grade—truly none (each was worth 0.5% of their total grade)—but the questions colored the students' impression of me as a teacher and of the quality of the course overall. They undermined students' trust and respect for me, and so damaged the learning environment. Those questions may not have hurt any student's grade, but they reduced my course grade. Students can be remarkably sensitive to what they perceive to be their teachers' attitudes and fairness.

Reinforcing that lesson, I know colleagues who don't circulate practice tests, leaving students surprised at questions they are facing. That feeds a perception that the professor doesn't care about them, which leads to students who perform poorly, have bad attitudes, and give bad evaluations. Everyone loses.

In setting expectations, remember that students have different skills. I try to avoid evaluation mechanisms that only reward one, and particularly the wrong one. I learned that lesson the hard way when I was a new assistant professor at the University of Alaska. My first class was general microbiology (with a lab). I spent lots of time developing the curriculum, but screwed up how I allocated the components of a student's overall grade. I assigned too much to the in-class exams and not enough to the labs.

Fairbanks had many older students, with a notable number of single mothers who were coming back to school to gain skills to get better jobs and to succeed independently. Those women were often among the best students—dedicated, hard-working, and committed. They knew exactly why they were there and who was paying their tuition. But equally, having been out of school for years, they were often not the best at taking tests. One woman in particular was one of those wonderful students who make teaching a joy; she was smart, thoughtful, and good in lab. But I think she only got a B- in the class, and it was my fault. Her exams brought her grade down, and I didn't feel I could break the contract about grading that I'd established in the syllabus. As a result of my errors, she didn't get the course grade I knew she deserved. I took that lesson, and restructured how I have organized every class I have ever taught since to more strongly emphasize take-home work (lab reports, problem sets, etc.) over in-class exams. That is more likely to reward students who are actually good students rather than just good test takers. And it more effectively maps grading onto the "stretched rubber band" model. Such activities require students to invest time and mental energy, not just in memorizing facts, but in truly learning the material and developing skills and understanding. By emphasizing these activities in grading, I try to weight the elements toward the right side of the flow in Figure 10.1. Exams are easy for us as professors, but students will learn more when we emphasize knowledge that persists, and focus less on facts that fade.

10.5 Pedagogy, Process, and Technology

Only after passion and communication are in place will specific classroom structures and technologies pay off. It's analogous to writing—if the story is boring or confusing, clever word play won't save the book! Similarly, strong acting can't salvage a desperate script. But bad acting can wreck Shakespeare.

Here, the stretched rubber band analogy is vital. Students learn when their brains are working, not just their note-taking hands. Traditionally, we expected students to work their brains at home when they studied the material we presented in class. That still happens, of course, but many students struggle to find time to study,[10] and there are other effective ways to help students engage their brains. This can start with something as simple as beginning a class with a question, one you will answer over the course of the class. It can involve classroom feedback tools such as iClickers, which allow you to pose questions during class and get instant responses from everyone. Not only does that push them to think, but it also lets you know whether they got the point. Such immediate feedback also helps you as the teacher. At the extreme, perhaps, is the flipped classroom where students are expected to study the material before class and then do activities in the classroom to test and solidify learning. Such approaches have the potential to transform teaching and can greatly enhance learning.

As with many specific topics I touch on in this book, there are other authors who can speak to pedagogical approaches and techniques way better than I can. I think I'm a good classroom teacher, and I get good (sometimes excellent) evaluations, but I get them more by relying on the first two elements of good teaching: passion and communication. And that, by itself, may be a useful message, particularly when you're juggling demands on your time.

---

[10] Many students have to work to pay for school. School is just one part of a busy and challenging schedule. Some also live in environments that are not conducive to studying.

But also, it may not be obvious how to distinguish among the elements. If you are clear in setting expectations and presenting the material, and you use effective pedagogical tools, students are likely to think you care about them. But, they'll probably be right!

Ultimately, to become a truly great teacher requires inborn talent, talent refined through years of hard work perfecting craft and skill, just as does winning an Olympic medal or a Nobel prize. But I don't aspire to greatness—I'll never be a Dr. Szell any more than I'll be an Einstein. But while greatness requires genes, goodness, even excellence, does not. Becoming a strong teacher calls for time, practice, and feedback from others—parallel to how we become strong scholars. All of us can develop our craft and become successful teachers. In fact, there is no excuse for not doing so.

Many of us, however, have jobs in which classroom teaching is one of multiple responsibilities. We must become successful teachers but without letting it undermine our scholarly careers. That is manageable. Care and let students see you do. Communicate effectively—remember that storytelling is at the heart of teaching, and that storytelling is a pretty straightforward art. Develop course structures to actively engage students' brains to encourage them to learn concepts and ways of thinking, rather than just memorize facts. If you can weave these pieces together, you will almost certainly become a successful teacher and may become of those standout memorable ones. But as with most things, success is more of a "Good-Fast-Cheap; Pick Two" model. You don't need to nail all three elements, but do aim to nail at least two.

SECTION 4
PROFESSIONAL COMMUNITIES

"If you want to go quickly, go alone. If you want to go far, go together"
African Proverb

Scholarly communities are global. We have colleagues who work in our areas across the nation and around the world. Journals and their editorial and review teams are drawn from everywhere. These systems are made up of human networks that bind together global research communities. Publishers and professional societies are the key entities in organizing and structuring our professional lives and activities. These global networks are vital to our careers as academics.[1]

11
Publishing Ecosystems

The Editorial and Review Process

Publish or Perish

"Publish or perish" actually *does* dominate our lives as faculty in research universities. Few of us, though, may feel that as the whip implied by the language. Rather, we have things we want to get out—ideas bubbling within waiting to be written. But whether we experience the need to publish as external pressure or internal drive, publishing means *publishers*.

We can post articles to the web and self-publish books, but our work gains quality, clarity, and credibility by the vetting, testing, and polishing associated with peer review and publication. Were we to eliminate journals and just post our manuscripts online (similar to physics and math's pre-print servers[1]), it would be up to each of us individually to sort out what of the mass of material was worth reading. In such an unregulated Wild West, systems inevitably develop to establish some certified "value" for papers, and to ensure they are archived in a form that the authors could no longer alter at will—that is, publishers!

At present, and for the foreseeable future, publishers remain gateways that control our careers, and when we start out in academe, they can be alien and scary. Editors and reviewers are more concepts than people: distant faceless, godlike entities weighing the fate of our manuscripts—and us! Or perhaps they're just a website to upload documents and from which to get decisions back.

But in fact, publishers, whether they produce journals or books, are just one more set of human systems that are part of our professional ecosystem. Even when a journal is *owned* by Elsevier, Wiley, or Springer-Nature, the editors who *run* it will be researchers who are active in the field—they are your colleagues and maybe your friends.[2] If you're starting out, they might be

---

[1] These fields use pre-print servers to share manuscripts, but still expect scholars to publish their articles in a journal to establish their validity and value (and to ensure archiving).

[2] Journals such as *Nature* and *Science* use full-time professional editors, but they are the exception. Most journal editors are working academics who at most get a modest stipend for their part-time service.

Your Future on the Faculty. Joshua Schimel, Oxford University Press. © Oxford University Press 2023.
DOI: 10.1093/oso/9780197608821.003.0011

people you want to postdoc with, or who are looking at your job applications! They are busy, overworked colleagues who do this as a professional service to support their academic community. The editors who work in book publishing are likely full-time professionals, but even they come to academic society meetings to engage with the community. I didn't only submit *Writing Science* (and then this book) to Oxford University Press because they're the world's largest, but because I'd met several of their editors and I *liked* them; they impressed me as people—and I knew that I hadn't a clue about publishing a book. But academic book publishers also rely on peer review in evaluating the proposals they receive.

When we start out in academe, we all have questions about how to navigate the publishing world. Many of the answers to those questions become fairly obvious when you recognize that the system you are really dealing with is *us*—busy people who do this work to support our community and the academic mission.

11.1 Journals versus Books

Publishing a book feels very different than publishing an article; I mean—it's a *book*. The idea of writing a book is terrifying. It's long and lonely work, and then you have to find someone who wants to publish it. But, whether we're talking articles or books, publishing remains a human system run by people. Editors are looking for pieces that have value to *readers*. Will readers access, value, and use the publication? But whereas articles go into the established structure of a journal, books each have to stand on their own.

Journal production is paid for either by subscriptions or author fees. Hence, the editors who handle manuscripts don't have to worry about finances.[3] A journal editor's job is to select the best and most appropriate articles given the journal's scope and target readership. The metric of an article's value, therefore, is *citation*—do people read and then cite the article in their work? By maintaining the quality of accepted articles, editors maintain the value of the journal. Journal editors have a relatively straightforward job that experienced scholars are well poised to do: assess the *academic* value of submitted articles.

---

[3] I'm Co-Editor-in-Chief of a leading journal, yet I have no idea what the journal's budget looks like!

A journal's finances and production are managed by the publisher. Whether we're considering a small department-run humanities journal, or a natural sciences journal that handles thousands of submissions each year, those costs typically run to several thousand dollars per published article.[4] Most of those costs, however, aren't associated with producing any particular article (turning manuscript files into a formatted article is largely automated), but with maintaining the information technology and management infrastructure.[5] Journals are information technology operations—they take in submissions, manage review and editorial decisions, support revision, and then produce and archive accepted articles. The battles over journal finances have been less about how much an article costs than about *who pays*: subscriptions paid for by libraries (free to publish for *authors*) versus article processing charges (APCs) paid by authors (open access for *readers*).

Books, of course, are different. A book has no pre-defined audience—each is unique. And the finances are not pre-established (sales can vary from a few hundred to tens of thousands), so the metric (at least for a publisher) of whether people value (i.e., read and use) the work isn't citation but *sales*. Will people buy the book? Or will only libraries and a few experts? Can it reach a wider community—if so, how wide?

Producing a book is expensive; on average, it costs a university press between $30,000 and 40,000 to publish a book (and some individual titles cost over $100,000 [Maron et al. 2016]). Also a greater share (ca. 70 percent) of that expense is associated with the individual book project, rather than with overall operations overheard. Financial concerns are therefore organic to a publisher's decisions—will it sell? As Germano (2013) notes, "There are not-for-profit and for-profit scholarly publishers; but there are no for-loss scholarly publishers."

Because recouping their investment is at the core of a book publisher's decisions, they can't rely on "amateur," part-time, academic editors. That is reinforced by the time it takes to produce a book: over a year from proposal to final volume with many production issues to deal with during the process. University presses may have boards of directors drawn from academic and library staffs, and they may use peer review processes in their decision-making, but their director and acquisitions editors are necessarily professionals.

---

[4] A colleague who manages a humanities journal estimates their per-article cost at ~$2,000, similar to a journal produced by a professional society.

[5] Hence the more selective a journal is, the higher its article processing charge (APC) must be because income comes only from accepted articles—journals don't charge submission fees.

11.2 Where to Submit?

The first question is always *where* to submit. How do you select a publisher? There is always pressure to submit to the highest profile outlet you reasonably can—but the keyword is *reasonably*. I'm always perplexed when I see submissions to *Soil Biology & Biochemistry* that are not about soil at all, or lack biology or biochemistry. Why would someone waste their time, and that of the editors, by submitting a paper that is so far out of scope that it's invisible over the horizon? I expect it's because SBB has one of the highest impact factors in its area, so authors figure "why not take a shot?" I understand trying that with *Nature*,[6] which publishes all areas of science. But with very few exceptions, journals have a defined focus that they describe on their home page. Read it. *And believe it.* If you are uncertain whether your work might be appropriate for the journal, contact the editor. Often a study produces results that can be used to tell a variety of stories, stories that might be most appropriate for different journals. Make sure you appropriately match your story with your target readership—send it to the right journal!

11.3 Preparing a Submission

After you decide which journal to submit your manuscript to, the next step is to prepare a cover letter describing the article, suggesting reviewers, and so on. Many people will tell you how important the cover letter is. As an editor, though, I ignore them (as do most colleagues I checked with). Why? And why the contrast to common advice? My reason comes down to the writer's adage of "show, don't tell." I don't trust what authors *tell* me in a cover letter; I trust what they *show* me in the actual manuscript. Cover letters argue why the work is wonderful and important. It may be, but I can't know that until I read the paper, so why not just go straight to the manuscript? Editors of other journals may operate differently, particularly the professional editors of journals such as *Nature* or *Science* who handle such a broad range of material they can't be knowledgeable in all of it.[7] In such cases, it makes sense that a précis on the paper might be useful. That is analogous to a book proposal, which lays out the vision, the need, the tentative plan, and offers sample chapters.

---

[6] Yup—been there, done that, and got the rejection letter.
[7] For example, *Nature* (as of 2020) lists only one editor on its Biology team who identifies their area of responsibility as ecology, and only one who targets evolutionary biology.

11.4 Recommending Reviewers

Being an editor is a lot of work—as a Chief Editor for *Soil Biology & Biochemistry* I handled two hundred manuscripts a year. Most needed at least two external peer reviewers, often through several revisions, and a substantial fraction of the submissions were in subfields in which I'm not deeply expert. So, I'm always grateful when authors recommend good reviewers. I rarely rely solely on that list, but I always start there for ideas and inspiration. Some authors provide suggestions that are really helpful, while others are useless. What's the difference? Consider the human perspectives—if you suggest the biggest names in the field, I already know them! And, I already know they are likely to decline the invitation. So if you suggest three big-wigs, you've given me no help at all.

Alternatively, when authors recommend reviewers who might appear to be too close, editors will worry that they won't provide a critical review—and so will likely ignore the recommendation. What is "too close?" As an editor, I try to avoid reviewers from the same university system (or even the same nation) as the authors unless I am confident I understand the institutional relationships. So I might invite someone from UC Berkeley to review a paper from UC Riverside, which I know are functionally independent. But if I'm unsure, I'll stay clear.

Those suggestions highlight what *not* to do. Instead, remember that this is an opportunity to get thoughtful reviews and to expand the community. So the ideal reviewers are people who've had the experience to develop vision and perspective, but still have the time in their lives to commit to doing a thoughtful review. That often translates to mid-career (e.g., associate professor), but assistant professors or even postdocs who've done interesting and insightful work also can be great reviewers. The perfect reviewer suggestion is one to which my response would be, "Of course! I hadn't thought of her, but she'd be great." Suggest several of those people, and you will make the editor happy.

11.5 After Submitting Your Manuscript

After you hit the "submit" button, there will be new sets of issues, such as how long to expect a journal to take before making a decision—that is, when do you consider sending the editor an inquiry about a paper's status? How do you deal with a decision once reviews come back? If you think the editor's decision is wrong, what do you do?[8] If you think a reviewer was prejudiced or otherwise inappropriate in their review, how do you deal with that? We all have had to deal with reviews and editorial decisions we thought were wrong or misjudged. But keep in mind that a journal isn't some faceless corporate entity—it's *us*. Your colleagues and peers.

Being an editor means you have to "play god," making decisions that affect people's careers. Yet, editorial teams are human and fallible—as a journal chief editor, I may play god, but I'm still just Josh.[9] We may make mistakes, we may misunderstand what you are trying to say, and we may even be momentarily careless or clueless. I'll plead guilty to all of those crimes. When we accept editorial roles we do so despite the huge amount of time and inevitable frustration, anger, and even anguish it can involve. But someone has to do it. And despite the frustration, these roles can be rewarding and satisfying; it's a way to pay back and pay forward, supporting peers and community, and advancing academe. Someone did it for us when we were starting out, and when the time comes, the way to repay is by taking on these roles ourselves.

Peer review isn't guaranteed to produce perfect outcomes. But it is akin to Winston Churchill's famous saying about democracy: "the worst form of government, except for all the others."[10] Peer review remains at the core of publication, and despite changes in publishing, it should remain fundamental. *Every* paper I've ever written has been improved by review and editing. I revere the process—even when it makes me want to scream.

As you interact with manuscript submission, review, and publication systems, keep in mind that a journal really is your peers. People whose good opinion you should value and whose time you should be sensitive to wasting. Remember the Golden Rule—treat them as you would wish to be treated yourself. With that simple perspective most answers about managing the submission and publication process should become obvious.

---

[8] Besides bitch to your friends—that goes without saying.
[9] Ask my wife!
[10] Churchill is usually credited with this, but his actual words were, "No one pretends that democracy is perfect or all-wise. Indeed it has been said that democracy is the worst form of Government except for all those other forms that have been tried from time to time."

11.6 Dealing with Editors

11.6.1 Is It OK to Contact the Editor with Questions about Reviews?

Will sending an inquiry to the editor reduce her total workload—namely, by addressing your question off-line, rather than in a resubmitted manuscript? If so, send the e-mail. It may take her some time to answer your questions, but if she is going to have to deal with a resubmitted manuscript, a quick inquiry may smooth the process and save a round of revision—or avoid a rejection. Dealing with a resubmission would certainly involve more work than answering your e-mail.

11.6.2 Is It OK to Submit a Rough Version to Get Input Before Polishing and Resubmitting?

I've heard it argued that because the average paper may be rejected at least once, then it makes sense to submit a rough version, get feedback, and then use that input to polish the draft for the next "real" submission.

But, won't submitting a rough version create extra work for the editors and reviewers? Duh! Of course. So no, it's *not* OK. It's *really* not OK! Submit the best version of your work you can, recognizing that it will still require revision. You should ask colleagues to do a friendly review before you submit, but that is collegial, *pre-submission,* review. Some fields, such as physics, use preprint servers where you can post papers and invite comment, but providing feedback is voluntary. When you submit a manuscript to a journal, the editors *must* deal with it. You will likely be a reviewer on other manuscripts, and may be an editor someday—would you want to have to deal with someone's "test run"? Nope? I didn't think so. So don't do it to others.

11.6.3 Is It OK to Submit a Manuscript to Multiple Journals at the Same Time?

Talk about making unnecessary work for your peer community! This is an utter and absolute no-no—in fact, it is a serious ethical violation invoking legal and copyright issues. If, as a reviewer, I got one manuscript from two journals, I would recommend rejecting both (regardless of its quality), flag the violation to the editor, and put the author on my "black list." Books, of course, are different; with a book, there is no legal commitment until you've signed a contract. Agents often contact multiple publishers, hoping to generate competing offers. But good luck on that!

11.7 Dealing with Decisions

Nothing is ever accepted as it was submitted. Reviewers always have suggestions. That might be because they feel a personal ego-driven need to put their fingerprint on the work. It might be because they feel a need to show the editor how good and smart they are. Simply saying "Great paper, publish it" might feel inadequate. But, no manuscript is perfect—there are always things that could be better, and a good reviewer will note those. An editor might well be suspicious of a cursory "accept as is" review—they might suspect the reviewer didn't read the work thoroughly.

Most authors don't submit a manuscript until they're confident that it is ready to publish, and when they're also pretty sick and tired of it, ready to put it in the past and move on to the next project. Dealing with reviews and revision can therefore be uncomfortable, dragging us back to revisit something we wanted to be done with.

We all can feel disgusted with reviews: Why didn't the reviewer get my point? I was clear! Were they careless or just stupid? But the main reason reviewers screw up is because the author did first! As authors, it is hard to avoid "The Curse of Knowledge." We know what we know, but reviewers don't know everything we know. Something that is obvious to an author may be opaque to a reviewer. Reviewers write reviews based on what they perceive in the manuscript and on what they take away from it. Thus, in my experience, when reviewers identify a problem, they are almost always right. Accept it. The real question is what the problem *is*. Reviewers might be right about there being a problem, but wrong about its nature. Is it in the methods and interpretation or just in the communication? If it's the latter, a reviewer may still perceive the problem as being with a method. In such a case, the solution wouldn't be to fix the method, but to fix the writing. You also need to explain to the editor, in your cover letter, how the reviewer had misidentified the problem.

As an editor, I may ignore cover letters on new submissions, but *not* on resubmissions. Those I look at closely; not so much at the line-by-line details, but absolutely at the discussion of the major conceptual points that either I, or the reviewers, had raised.

You can argue with reviewers, and the cover letter is the right place to do that. Why do you think they're wrong? Why are you are going to solve a problem some other way? But make the changes in the paper as well to help readers avoid that reviewer's misunderstanding. After all, no one—besides the editor—will ever read your beautiful compelling explanation in the cover letter.

There is a human dimension to responding to reviews as well: Reviewers and editors want to see that authors are responsive to the input and to the process. If you push back on too many things, editors may be suspicious that you are blowing them off, and the reviewer will likely be doubly critical if they feel you ignored their concerns.

11.8 Rejection

We all experience rejection. It sucks. Yet, journals reject papers and publishers reject book proposals. Get used to it. It's not a question of whether we will have works rejected, but how we deal with it when they are. With journal articles, when you get a rejection letter, it's easy to get upset at the editor who signed the letter, but the decision likely began with the reviewers whose criticisms underlay the editor's decision. So you can either send the work to a different journal or appeal the editor's decision. Alternatively, you could just put the manuscript in a drawer and forget it.

Assuming you go with the first option and resubmit somewhere else, you still have to decide how to address the reviewer's criticisms. You could decide they were invalid and ignore them, resubmitting the manuscript unchanged to a different journal. That's not wise, because it's unlikely that there is nothing of value in those comments. You may disagree with them, they may be snarky and ill-tempered, and they may be wrong, but it is still likely that the reviewer was trying to follow my reviewers motto: "Friends don't let friends publish bullshit" and were trying to be constructive. Use that input to polish your work further. And sometimes, their recommendation to reject is valid—a flaw you might have overlooked? Then why would you want that paper published? I reviewed a paper once that had a deeply and utterly fatal flaw—the results and conclusions were pure artifact, resulting from an error in how the experiment was done. It was rejected, and the authors should have thrown it away. Instead they got it published elsewhere—and not, as it deserved, as fiction!

Have I forgotten who the authors are? Nope. That was an extreme case, but reviewers have long memories, and we might be asked to review your work in the future—or to write a promotion letter! Treat the review community with respect.

Alternatively, you might be inclined to go with Option 2 and appeal the rejection decision on the basis that it was inappropriate. Either the reviewers missed the point, or perhaps their negative review was based in a personal animus toward you. Both do happen. Reviewers are human, and humans screw up. We also can be prejudiced, pissy, and peevish; we have bad days. It's possible that a reviewer has something personal against you, but it's more likely a reviewer (or editor) just blew it. We've all seen those reviews, and I suspect we've all written them. To err is human. So if you think a decision is misguided or inappropriate, call the editor on it. As an editor for *Soil Biology & Biochemistry*, I get several appeals a year, and I've appealed a few rejections in my life. One of those was to the editor of *Nature* who'd rejected Jeff Chamber's paper on the age of trees in the Amazon rainforest (Chambers et al. 1998)—they don't make annual rings so no one knew. The editor responded to the appeal, "I don't know what I was thinking," and he ultimately accepted the paper. I am still in awe that he was so open and forthright about having made a mistake—a "Gold Star" moment in editing.

When you get a disappointing or an inappropriate review, however, remember that the idiot of a reviewer was chosen by the editor, and that the editor is your colleague. So when you get one of those decisions, before you blast off an angry reply, let it sit for three days to cool down. Then consider that the problem may not have been with the reviewer, but with what they were reviewing—*your* paper. Did he misunderstand because he's an idiot or because you were unclear? It's unlikely that it was 100% the former.

If you appeal a decision, at least try to be considerate and respectful. Get an outside, impartial, reader to check your letter—does it come across as snarky or respectful? Acknowledge that there may have been problems with the paper that may have led the reviewer astray and note how you could fix those. Dealing with a reasonable appeal is fully within an editor's responsibility—if I screwed up, I'll reconsider and fix it. But if I have to deal with an author who is in a huff, it causes extra work and headaches I don't deserve for just trying to do the best job I can. First, I have to convince myself not to react with my initial inclination—to just say "F-off!" Then, I have to sort through what may be valid argument rather than just peeve. Being nasty is also more likely to motivate an editor to justify their decision, rather than to reconsider it. Editors may be human, but the role forces them to make decisions and act as god. In appealing a decision, you're asking a colleague to do you the courtesy of reconsidering their decision. So do it nicely.

11.9 Books

Acquisitions editors are always on the lookout for projects that have both intellectual value and marketability:

> [T]here are always hunter-gatherers at scholarly houses who want the exceptional dissertation. Of course, editors pay particular attention to award-winning dissertations in their commissioning fields. But word of mouth is still the editor's secret weapon. Every successful scholarly editor relies on a network. Trusted faculty advisors can identify the most promising dissertations being written in the discipline, and if you're writing one of them an energetic editor may be in touch with you before you've hammered out Chapter 3. (Germano 2013)

Equally, of course, most publishers won't even look at the unrevised dissertation itself—dissertations and books are different beasts. Germano notes four aspects that often need to be addressed in revising a dissertation into a book: audience, voice, structure, and length. A dissertation is to prove yourself to your professional community; a book is to sell a bigger story to a wider community. The voice and structure of a dissertation are often formulaic and *academic*; they also are often too long. Many excellent books have grown out of dissertations, but revising a dissertation into a book is a grand challenge for an assistant professor. It requires adapting the core into a form that will engage more readers than just your Ph.D. committee.

Acquisitions editors may seem like Smaug guarding the dwarven treasure in the Hobbit—an undefeatable defender of the ultimate treasure; but they, too, are humans who rely on you to provide that treasure. And that doesn't mean chewing you up and spitting out your cracked bones. They need our content, just as we need their production and marketing systems. Building relationships with editors is at the heart of becoming successful in this arena. And this goes both ways: editors and publishers recognize that we don't get trained in how to write books in graduate school, and that editors have a role in "training scholars to be published authors" (Marron 2016). First-time authors take a lot of editorial support and hand-holding as they develop and learn how to adapt scholarly work into viable book projects. I can vouch for that. I had (and still have) very little clue about how to approach a book; I have relied on the editorial team to guide me through the process. I still have a long collection of e-mails to and from the production team who worked on *Writing Science* in getting it through copyediting and production; if you're reading this, I likely have an equally long collection related to this book. There's a huge amount that I'd never dreamed about but which had to be worked through.

Unsurprisingly, therefore, Marron and colleagues (2016) note how much the author can drive up the cost of producing a book: "When staff allocated additional time to working on a particular book, it was often because they could recall—often in some detail—the difficulties of working with a specific author on a specific project." When authors don't follow instructions, ignore format requirements, miss deadlines, and so on, they waste staff time and increase costs. Even worse, of course, is when authors try to make changes once the manuscript is already in production. With my books, I got explicit and clear marching orders: "no changes to the proofs that would change pagination" because that would screw up indexing, disrupt publication schedules, and drive up production costs. So don't do those things! The core rule from Chapter 7 applies here: "Take care of the people who take care of you." The publisher's team becomes *your* team.

Book publishers are increasingly facing challenges analogous to those that journals have faced over the last decade or more, with debates about traditional publishing versus author-driven "open-access" production. Such debates ultimately come down to the functions publishers provide. A publisher doesn't only convert word processor files into journal articles or books. A publisher also curates and archives the material—they manage the evaluation, review, selection, polishing, and "author training and development." That has long been a part of academic publication, and it is one I value. I dread having to filter through the raw manuscripts people might upload to figure out which of them was actually worth reading. As a journal editor, I already spend enough time filtering and separating the "wheat from the chaff" so that when someone looks at a paper published in *Soil Biology & Biochemistry*, they can rely on it being "wheat." I respect my colleagues who do the same service at other journals. I value the polishing involved in the review and revision process. I'd hate to see the end of those services.

The bottom line of this chapter is to recognize that publishers comprise just one more human system that is fundamental to our success as academics and to the overall functioning of academia. Regardless of concerns over the finances in academic publishing, as academics, we don't deal with the corporate executives, but rather with the editorial teams who handle our work. They either are us—working academics—or at least they work closely with *us*. Publishing successfully and effectively means understanding publishing and editorial systems. That will set you up to work with the editorial teams and to create the most successful and powerful publications you can.

12
Who Put the *Peer* in Peer Review
Being Part of the System

Friends don't let friends publish bullshit.

The decisions that most powerfully influence our careers almost all rely on peer review. Whether that decision is to accept your manuscript for publication, to fund your proposal, or even to grant you tenure, peer review is organic to the process. When we start as graduate students, we aren't yet "peers" and so we are almost entirely on the receiving end of review, but as our careers develop, we can increasingly expect to be invited to participate. Every system I know needs people to review, so as soon as you gain any name recognition, from publications, presentations, or casual connections, you will be asked to review manuscripts and proposals.

Reviewing the work of others—your peers—is a fundamental responsibility as an academic. Editors and program officers are senior members of your field; they are the people who handle your papers and who might be asked to write letters for a tenure case. Doing your part to keep the review systems functioning is thus an obligation—do for others what they do for you—but also a way to establish yourself. Contributing to your profession's infrastructure and operating systems is part of being a professional.

We rely on peer review because research is hard. It doesn't matter whether you're figuring out how a viral coat protein unfolds to enable the virus to penetrate our cells, why the stock market bubbles, or why the French Revolution exploded into the Reign of Terror; nature and human societies are complex, so scholarship always involves judgment and interpretation.

Communicating our research is also hard. Not only is the material itself hard, but the "Curse of Knowledge" makes communicating it even harder. As authors, we know what we know, and we know what we mean. But we can't know what all our readers know or think, and we can't know how they will interpret our words. The ancestors of *Homo sapiens* began developing spoken language over 1 million years ago (Morgan et al. 2015); it is part of what defines us as a species. But writing is unnatural (Dryer 2016), having only appeared about 5,000 years ago. A word's meaning may depend on its context, a context that in spoken language we can highlight and clarify with facial expressions, vocal tones, and body language; we also have "error-checking" tools in spoken language (Enfield 2017). To interpret a word in written language, we have none of that; we must rely on a word's immediate context—the sentence and paragraph it appears in—and on community norms and expectations. For example, does "succession" mean the replacement of one plant community by another, or the replacement of a king by his heir?

Expressing complex ideas calls for careful and skilled writing. But, of course, most academics aren't great writers. If you're like me, in school you focused on the content in your classes and thought about the information and the ideas, rather than on how to communicate them. I certainly figured my literature teachers had little to offer me. But if you thought as I did, you'd have been just as wrong as I was—we all have a lot to learn when we start writing professional work. "Writing" isn't just putting words into sentences and paragraphs, but crafting documents to change how your readers view and understand their world.

Combine these issues, and you can see why academic writing is so hard. So hard that no one ever gets their manuscripts completely right with the first submission. No one. Ever.[1] I may be a good scientist and a good writer, yet I rely on peer review to identify issues I hadn't thought of or that were unclear in my writing. Some of my papers were only slightly polished by review, others were transformed. Peer review is fundamental and essential. Participating in the process is therefore one of our core professional responsibilities.

That responsibility is one reason I will also argue, strongly, that you should respond to *every* review request—and respond quickly. It's fine to say "no," but do it immediately. An editor sends out several requests hoping to get enough "yeses" back, but they don't want to send out so many that they end up wasting people's time getting more than they need. After sending an invitation out, they have to wait to hear back before sending out more. It takes but a minute or two to decline, and then the editor can go down their list. When people don't respond at all, the request hangs out there for a week or so before the editor can assume your answer was no.

We all hate it when the review process takes forever. So help—at least respond to the request! And if you agree, get the review done in reasonable time. Don't be one of those people who agrees to do a review but then flakes and is very late, leaving the editor debating whether to send one more reminder, or to just abandon hope and start afresh with a new reviewer. But it could

---

[1] A paper being accepted as submitted is like a Unicorn: a myth. I've certainly never seen one.

take weeks to get someone new, who would then have several weeks to do their review, and they might also run late. If you want your papers reviewed quickly, return the courtesy: respond quickly to requests, and get the review done on time (or at least close to it). If you really hit a crisis where you can't scrounge up the time to do a review you'd agreed to, contact the editor and let them know. Peers and colleagues deserve that much consideration. And keep in mind as well, editors will remember who contributes well versus who is always late and then sends in skimpy, minimal reviews.

12.1 What is Peer Review

The key thing to remember about peer review is that it is not something we do *to* each other—it's something we do *for* each other. The important word is *peer*. It's a professional service and a professional courtesy. Thus, when approaching reviews, my governing motto is:

> *Friends don't let friends publish bullshit.*

The vulgarity is not to suggest that my friends submit bullshit, but to hammer the point that we serve our friends by helping make their work as strong as it can be. Published is *forever*, so if a paper really is flawed, I shouldn't want it published. One more paper on my CV might look good to my departmental colleagues when they evaluate my merit and promotion review, but the last thing I should want is my disciplinary colleagues to ask, "I wonder who reviewed that last paper of Josh's—how did that piece of junk get published?" If that were the case, *kill it*, please. Ultimately our careers are defined by our best work, not just its quantity, and by what our peers think of that work and of us. We all write a lot of "fill in the blank" papers that are more incremental than transformative, but people who become leaders do so by publishing work that changes how we think and how we look at problems. One really good publication can do more for your career than a bunch of just-OK ones. That's why I argued in *Writing Science* that "You don't succeed as a scientist by getting papers *published*. You succeed as a scientist by getting them *cited*." The positive, nonvulgar, way to put my "friends don't let friends . . ." statement, therefore, would be:

> *Friends help friends succeed.*[2]

---

[2] You can see why I like the "bullshit" version better. This version, though true, reads like a weak platitude.

When you are asked to review a manuscript, remember that your jobs are to: (1) support the *editor* in making a good decision on the piece's fate, (2) support the *authors* by giving them advice on how to make the paper as strong as it can be, and (3) support your professional *community* by ensuring the overall quality of the body of published work and of the peer review process overall.

Our goal should be to ensure that the best work gets published and that what gets published is its best. That means being supportive when appropriate, but equally, it means being critical when that is called for. That's equally true whether the authors are old friends or people we've never met. Friends help friends succeed, even when that can be painful.

The most important and highly cited paper I will likely ever publish (Schimel and Bennett 2004) was a conceptual synthesis in the journal *Ecology*. The editor handling it, Peter Groffman, is a friend I've known since graduate school. Peter held our feet to the fire and ran us through the wringer over that paper! He didn't insist that we do things the way he might have, but he pushed us to address his and the reviewers' concerns through two substantial revisions. As I submitted the final version, I thought to myself, "If Peter accepts this one, I owe him a beer—if he sends it back for more changes, I'm going to pour it on him." Peter knew the paper was an important contribution, and I am grateful for the work he did to enhance it—it was journal editing at its finest.

Another case of "friends help friends" was when an editor sent me one of Dr. Elisabeth Holland's papers to review (back in the hardcopy days). Beth wasn't just a dear friend—she was my sister-in-law. Talk about conflict-of-interest! Of course I couldn't review the paper officially, but I read it and prepared a critique, which I sent her directly. She told me it was the toughest review she received. I love Beth dearly; of course I wanted to help her publish the best paper possible.

Treat everyone that way: This is a service we do for our peers, colleagues, and friends, those who are the authors, *and* those who might be the readers. Our entire system of scholarship depends on peer review, and our willingness to be our friends' and colleagues' critics.

12.2 How to Write a Good Manuscript Review

A "good" review is an analysis that is useful and constructive for both the editor and the authors. It helps the *editor* decide whether a paper should be published, and which changes they should request or require. It helps the *author* by offering guidance on how to improve their work so that it is clearer and more compelling.

As you approach a review, keep in mind that manuscript review isn't just criticism—it's triage. Triage, in this context, comes from military medicine. When wounded soldiers are brought into an aid station, busy doctors must separate those likely to die (regardless of what surgeons might do) from those who can be saved by appropriate medical care. All manuscripts come into journal offices as "wounded soldiers." Some of those manuscripts only need a light bandage, others require major surgery, but they *all* need some editorial care.

When a paper is submitted, the editor and reviewers must do triage: Does this paper stand a chance of becoming "healthy" and publishable? Or is it so badly "wounded"—suffering from a poor study design, inadequate analysis, or a weak story—that it should be allowed to die in peace (i.e., be rejected)? An editor at a top-tier journal such as *Nature* is like a surgeon in the swirl of desperate battle, getting a stream of patients that overload any ability to treat them all; a high proportion must be rejected and allowed to die. At a specialist journal, the flood of manuscripts is less, and so we can "treat," and eventually publish, a greater proportion of the papers.

Typically, an editor makes a first triage cut—if the paper either doesn't fit the journal's scope or is so badly off that it obviously has no chance of surviving, the editor will reject the paper without getting external reviews.

But triage doesn't end with the editor. When you receive a manuscript to review, the first thing to address is the triage question: Is this salvageable? Can it reach a level of "health" appropriate to publish following a reasonable investment of time and energy on the part of the editorial and review team? A manuscript may have a data set that is fundamentally interesting and could support a good paper, but still have an analysis or discussion that is in such poor shape that it would be best to decline the paper and invest limited editorial energy elsewhere.

When you write a review, the first paragraph(s) therefore should target the triage issue; it should frame your argument for whether the paper should be rejected or should move forward into revision and polishing. First, is the core scholarship sound, interesting, and important enough for the journal? Second, is the manuscript written and argued well enough that, with a reasonable level of revision, it will likely become publishable? If the answer to either of those questions is "no" then you should recommend that the editor reject the paper. Explain your reasoning and analysis clearly and objectively enough that the editors and authors can understand your recommendation.

If you answer "yes" to both questions—the work is sound and the paper constructed well enough to be worth fixing—you move from diagnosis to treatment, and the audience for your review shifts from the editor to the authors. Everything, now, should focus on helping the authors make their paper *better*. That doesn't mean avoiding criticism, but criticism should be linked to discussing how to fix the problem. Going back to Chapter 8 on mentoring, this is how we are kind to our colleagues—by helping them.

Your review should identify where you think the authors were unclear or wrong in their presentations and interpretations, and it should offer suggestions on how to solve the problems. The tone should be constructive and fundamentally supportive. You've decided to recommend that the "patient" be saved, or at least may be savable, so now you're identifying the "wounds" that should be sutured or bandaged. It doesn't help to keep beating a paper with its limitations and flaws, unless you are going to suggest how to fix them! If the problems are so severe that you can't see a solution, why haven't you argued to reject the paper?

In this section of a review, you are free to identify as many issues as you wish—but be specific. If you say, "This paragraph is unclear, rewrite it," that won't help an author—if they could see why you thought the paragraph was unclear, they probably would have written it differently in the beginning! Instead say, "This is unclear—do you mean X or do you mean Y?" If you disagree with the logic of an argument, lay out where you see the failing, why you think it fails, and ideally, what you think a stronger argument might look like.

It is easy to fall into the "Curse of Knowledge" when we write reviews because we don't generally edit our words as carefully as we do when we write papers—in fact, we rarely do multiple drafts of reviews. You know what you're trying to say, so it may be obvious to you, but your message is never as clear to readers. Because authors and reviewers both suffer curse of knowledge problems, it's easy to get caught up in a cursed cycle. The author is unclear, but a reviewer is unclear about what is unclear, leaving the author flailing trying to figure out how to fix it! A good review needs to be clear and concrete, just as does a good paper.

Remember that it is not a reviewer's job to rewrite the paper—it's still the authors' paper. If you don't like how the authors phrased something, you can suggest changes, but you are trying to *help*, not *replace*, the authors. If the disagreement comes down to a matter of preference, rather than of correctness or clarity, it's the authors' call. If the authors aren't taking their story to the level that you see it could reach, at some point you still have to acknowledge that the paper reached "publishable," and let it go.

When I do a review, I usually make side notes and comments as I read. Then I collect these thoughts, synthesize my critical points about the intellectual framing of the paper, and write the guts of the review—the overall assessment. I target that discussion toward the editor, since my first responsibility is to help her with triage. She will ultimately make the accept/reject decision and she will tell the authors what issues to address for the paper to become acceptable. Then, I include my line-by-line specific comments. Those are aimed at the authors, as they are about the details of the paper. These typically run from half a page to a few pages of text.

Sometimes my reviews get longer—I've written six-page reviews, reviews where I wanted to say that I thought the paper was fundamentally interesting and important, but that I disagreed with some important parts of it and that I wanted to argue with the authors about those pieces. I typically sign those reviews because (a) I figure it will likely be obvious who wrote it, and (b) I am willing to open the discussion with the authors: It isn't an issue of right versus wrong, but of opinion, and I think that the science might be best advanced by having the debate.[3]

12.3 How to Offer a Specific Recommendation?

Journals typically ask you to select a single overall recommendation. That is almost always a check box on a web form. These fall into several general categories, ranging from accept to outright reject, although journals use different versions of the specific terms.

Accept: This means the paper is ready to publish as is. I don't think I've ever used this on a first review.

Accept following minor revision: This means that the paper is conceptually sound and just needs some polishing. It doesn't need a "follow-up visit"—that is, you don't think it will need re-review, and you are telling the editor that *they* should check the edits and probably accept the next, polished version.

Reconsider following revision: The paper is wounded, but savable. The problems go beyond clarity or minor edits; the paper requires some rethinking. It will therefore likely need re-review. Some journals call this "reject but resubmit" or something similar. There is no promise that the

---

[3] In one case, the conversation that grew from that signed review led to a new collaboration, several coauthored papers, and a long-term friendship.

paper will ultimately be accepted (although that may be likely—as an editor, I never use this decision unless I hope to be able to ultimately accept the paper). If you recommend "reconsider," I hope you will also agree to do that re-review.

Reject: The paper should be allowed to die. Either it is fatally flawed in its core, or the story is so poorly written that it is not worth the editorial team's investment to try to make it publishable. Rejection usually means the authors may not resubmit the paper unless they appeal the decision and get permission from the editor.

Keep in mind that as a reviewer, you are typically anonymous. The editor is not. If there really are deep flaws in a work, give the editor cover by recommending "reject." If they choose not to take your advice, they get to be the good guy but still to push the authors: "Reviewer #1 suggested declining the paper, but I think you might be able to solve the problems, so I'll give you a chance to try." That, of course, carries the implication of, "but if you don't, I *will* reject it." If instead, you are nice and recommend "reconsider," that leaves it to the editor—if they decide to reject, it's all on them and they're the villain. Those of us who have served as editors signed on to do that job, and we will, but we appreciate your help. Give your most honest and accurate assessment, but remember that the editor attaches their name to their decision.

Remember also that you are *advising* the editor. You are not deciding whether to reject a paper—that is the editor's job. I've been amused when reviews come in with language such as "I reject this paper." There are journals where decisions may feel like the editor is just counting votes, rather than assimilating input. But with good journals, particularly those published by professional societies, editors are entrusted by their community to manage the process and ensure that it remains robust and supportive to all involved— that means synthesizing the input and applying their judgment to make their best decision.

12.4 Reviewing Revisions

How does this advice change if you are getting a revised manuscript back for re-review? The same principles apply of course. Your job is to help the editors with their decision, and the authors with their paper. Yet, reviewers sometimes become possessive and insistent about their suggestions—for example, getting annoyed when authors don't do exactly what they had recommended.

Don't. First, remember that the editor likely received two or three external reviews that might have varied in their assessments and recommendations. Editors need to synthesize all that input before making a decision and offering guidance, or marching orders, to the authors. Then, authors might have different ideas about how to solve the problems and address reviewers' concerns. As long as the authors' solution works, *it works*. When doing a re-review, your job is to determine whether the paper has crossed the threshold of acceptability, not whether the authors have done everything you had suggested, and particularly not whether they did everything in way you might have suggested. I have accepted papers that I know could still be better, but it's not my paper. In the triage model, the question is not whether the patient is 100 percent healed, but are they healthy enough to release from the hospital.

The more difficult call is when a paper has improved, but not enough—it hasn't crossed the threshold of *acceptability*. I expect a paper that starts at "reconsider" to step up to "minor revisions" *en route* to "accept." But what if the paper still needs additional major revisions before it closes in on acceptability? The paper might have gotten better, but not enough, and the trajectory is relatively flat. In such a case, you should probably recommend rejecting the paper. It's not that the paper can't become publishable, but having given the authors advice on how to improve it, they either chose not to take the advice or couldn't see how to. You can't write the paper for them and you can't force the issue—we all have finite time and energy to invest in a patient that isn't getting better. At some point, we just have to make the hard call, move them out of the hospital ward, say "I'm sorry," and let them go.

12.5 To Sign or Not to Sign

Increasingly there are review systems that involve disclosing reviewer identities. But, should you sign your reviews? You should certainly write every review with such a level of professionalism that you would be willing to stand behind your comments. I have mixed feelings, however, about signing reviews. I went through a phase where I signed every review, believing that it was the honorable thing to do. But then I faced a paper that was bad enough that my review had to be very critical—harsh enough toward people I know that I didn't want to put my name on it. I realized that if I was only going to sign positive reviews, I wasn't being honorable, but rather asking for kudos and thanks. So I stopped signing reviews as a routine.

The key thing is that a review be honest—it *must* provide the editor with insight to support their decision and it *must* provide the author with advice on how to improve the paper, or explanation for why you think it can't be fixed. If remaining anonymous makes it easier for you to be as direct and honest as you should be, then remain anonymous. Do not, however, hide behind anonymity to behave badly, be rude, or be inappropriately critical. Reviewing is a service that we do to support our communities, not an opportunity to stick a knife in the back of a competitor.

These days, the only reviews I sign are those where (a) I think it would be obvious that I was the reviewer and (b) where I am inviting a broader discussion. That means I sign at most a few reviews a year.

Reviewing is both professional service and obligation—it's what we do for one another to advance our scholarship. We help our colleagues by identifying areas where the work is unclear or the arguments weak. Review can be painful, but writing is hard. Accept it. We all rely on peer review, so embrace the process when you're being reviewed, and do the best job you can when you are the reviewer.

12.6 Proposal Reviews: A Different Beast Entirely

Funders are a fundamental part of the academic ecosystem; they include both internal university programs as well as external programs such as nongovernmental organizations (NGO's) and State or Federal agencies. Even scholars who are in fields where one can function without external grant funding do better with resources to at least hire student assistants.[4] Almost all those programs rely on peer reviews when they make decisions, and so contributing to them through review is part of supporting your community. Yet, peer review is different when you're reviewing proposals as opposed to papers.

In the triage model, the most important part of a manuscript review is *treatment*—make the "patient" healthy, if possible. Proposal reviewing is different: *there's no treatment*. It's pure diagnosis and evaluation. It is still community service, given that we have a vested interest in funders making good decisions. But when we review proposals, the Principal Investigator (PI)[5] is

---

[4] For example, many humanists, but some natural scientists as well, emphasize theory over experimentation.

[5] Few things bug me more than when a someone calls a PI a "Principle Investigator." A "principle investigator" investigates principles: a philosopher, not a researcher! A PI is the lead, principal, investigator.

not the audience—the review panel and program officers are. We're not trying to tell PI's how to make their work better, we're telling the agency whether they should fund them. The PI's see reviews, but that is almost like spying—looking from the outside at what went into the decision.

When you review a research proposal, remember that proposals are works of fiction—the PI's are not going to do exactly what they wrote. A proposal isn't a promise, but a plan, and the military maxim, "no plan survives contact with the enemy," applies. The researchers may have great ideas, but nature won't cooperate, or they'll recruit a student or postdoc who takes the work in different directions. That's research. You still must aim to achieve the project's core goals, but the plan *will* mutate. If you knew enough to describe exactly what you will do over three years, you knew enough that you didn't need to do it! We do the research because we don't know the answers. We rely on PI's to use their judgment to sort out glitches that arise.

To recommend funding a proposal, therefore, it should be pretty awesome; awesome enough that you have confidence that (a) it is worth doing, (b) enough of it will likely work, and (c) the PI's will be able to work around the parts that don't work and still achieve their goals. If major elements are likely to fail, or you lack confidence the investigators will be able to solve the problems that arise, you should say so and recommend rejecting it. When you review a proposal, therefore, you must answer two questions: (1) Is the proposal exciting and novel enough to be worth investing limited resources? (2) Is the proposal technically sound enough to be doable?

A PI shows the novelty of their questions by defining the knowledge gap and the boundaries of knowledge (not just a fuzzy "little is known about this") and by framing clear, answerable questions, and/or falsifiable hypotheses (not fluff such as "increasing temperature will *alter* the structure of communities," but how it will alter them.). PI's show the work will likely succeed by laying out the study design, discussing methods in appropriate detail, describing how they will address risks, alternative strategies in case pieces don't work, and so on.

Because not all proposals can be funded, reviewing is inherently relative: How does this stack up against the competition? You aren't reading all those, so you have to assume a baseline of comparison. That is why the first proposal I ever reviewed took me several days; now it sometimes only takes an hour. I had to develop a reference for what a good proposal looks like—the job gets easier the more you review. And, of course, the baseline differs among funders: some National Science Foundation (NSF) programs fund only 10 percent of the proposals submitted, while our campus Academic Senate funds (at least partially) most of the proposals they receive.

12.7 Review Scores

Part of most reviews involves giving an overall score to the proposal, analogous to how we grade student papers. The official definitions we assign these scores, however, rarely match reality. For example, the NSF official definitions are:

Excellent: Outstanding proposal in all respects; deserves highest priority for support.
Very Good: High-quality proposal in nearly all respects; should be supported if at all possible.
Good: A quality proposal, worthy of support.
Fair: Proposal lacking in one or more critical aspects; key issues need to be addressed.
Poor: Proposal has serious deficiencies.

Over the years, I've developed my own definitions that I believe more closely match how the scores work in practice:

Excellent: This is a *very good* proposal that deserves funding. Exciting questions and it has no major flaws. If I'm on the panel, I'll fight to see this goes into "High Priority."
Very Good: This is a *good* proposal. The questions are interesting, but they don't blow me away, and there are likely some minor gaps. Functionally, this is a neutral score, not really arguing strongly either way. On a panel, I'd say positive things, and would be happy if it were ultimately funded, but I wouldn't fight for it.
Good: This is a *fair* proposal; the ideas are valid but not exciting and/or the approaches are weak, but not fatally so. This might produce solid research, but I don't think it's competitive. If I score something *good*, I'm deliberately damning it with faint praise.
Fair: This is a *poor* proposal. It should *not* be funded. There are major gaps in the conceptual framing, weaknesses in the methods, and/or it lacks novelty.
Poor: This score is not really for the program officer, but for the PI. Giving a *poor* is deliberately and actively mean; it's twisting the knife of an already lethal review. It says: *I want you to hurt for wasting my time with this piece of junk!* In my entire career, I've rated only a few proposals as *poor*, and never to a junior investigator. I've read horrible proposals by new investigators

who just haven't figured it out, but still rated them *fair*. Nope, *poor* is reserved for people who have no excuse and should know better.

To write a useful proposal review, remember you are really just making a recommendation (fund vs. don't fund) and then providing justification for that recommendation. If you think a proposal is super, why? What is novel? Why is the project so clever? Why is the inclusiveness part more than just, "we'll recruit underrepresented students from our local community college." How have the PI's woven the outreach activities into the research? As an *ad hoc* reviewer, use your expertise to argue to the panel what they should *recommend*. If you're a panelist, give the program officer the information and rationale they need to help them *decide*. Do those things well, and your reviews will be useful and appreciated.

Editors and program officers are part of our scholarly communities. They carry out critical work on which we all depend. Thus, as academics, it's our responsibility to support them in their jobs so that they can support us in ours. Once again, this distills down to *take care of the people who take care of you.*

13

Professional Communities

The more privilege you have, the more opportunity you have. The more opportunity you have, the more responsibility you have.

Noam Chomsky

Remember who your real peers are.

David Schimel

As academics, our day-to-day lives revolve around our departments: classes, meetings, students, chance conversations in the hallways. Our department is the family we are "born" into as an assistant professor; but as with our biological family, we may have a beloved sibling, but equally a crazy uncle who can be counted on to blurt out impolitic comments in gatherings, a crotchety aunt who mostly sits in the corner and says little but has a sharp tongue, and a young cousin who still seems clueless and bounces like a puppy. We live with these people, but we did not choose them, and most aren't our research collaborators.[1] Your campus may provide a somewhat larger group of scholarly colleagues, given that department boundaries are porous—is biochemistry biology or chemistry? Is studying French literature French or literature? As a soil/ecosystem ecologist, for example, I have a few collaborators in the Geography Department and in the Bren School of Environmental Science & Management.[2]

But then we go to conferences where there are hundreds of people to whose work we pay attention. I'm a member of the Soil Ecology Society, which is a small, focused group whose biennial meetings still draw several hundred people. Large societies are way more impressive: The American Geophysical Union (AGU) has more than 60,000 members, and its annual meetings draw over 10,000 participants.[3] But AGU encompasses an array of

---

[1] I've only meaningfully coauthored papers with two department members out of more than fifty who have served in the department, plus one more where I contributed to a colleague's postdoc project.

[2] I've coauthored papers with a total of eight University of California Santa Barbara faculty—one-quarter of my publications. In contrast, one-third are coauthored entirely with off-campus collaborators. The rest are sole authored or only with my students and postdocs.

[3] And AGU is dwarfed by the American Chemical Society, which has over 150,000 members.

subcommunities, with twenty-five sections as diverse as Space Physics and Hydrology. The American Historical Association is similar; with over 12,000 members, it identifies more than 120 Affiliated Societies, each of which has a specialized focus, such as Legal History, History of Science, Military History, or even Automotive History. Each of these smaller groups has from several hundred to a few thousand members. That is probably not an uncommon size for a "working community." There are, after all, over 430 Ph.D. granting universities in the United States; there is likely someone with a specialty similar to yours in many of them, particularly in the approximately 150 universities that grant more than one hundred Ph.D.'s a year (NCSES 2019).

It is in these dispersed communities that we live our intellectual and scholarly lives. I only came to understand that, however, when I was an assistant professor. I was at the business meeting of the Soil Biology section of the Soil Science Society of America, and I looked at who was speaking regularly: a bunch of us who'd first met when we were graduate students and postdocs in the 1980s and became assistant professors around the same time. I had the terrifying epiphany that although we might then be the "Young Turks," we were also the incipient "Old Boy's Club"![4] These were the people I was going to grow old with and would be hanging out with at conferences until we retired. And that's true—those people are still among my closer friends and colleagues, they have handled my manuscripts as journal editors, sent me students and postdocs, and recruited my students in turn.

These professional communities comprise our *chosen* friends and colleagues (even if you don't necessarily like all of them). They are the intellectual communities that frame and define the long-term sweep and direction of our careers. They are where we present our work, publish our papers and books, and where we live as scholars. These are the people who review our proposals, papers, and promotion cases. When department politics are frustrating and toxic, we look for friendship and support in these wider networks. That is why I repeat my brother's advice as a lead quote for this chapter: *Remember who your real peers are.*

The global community is also the key to the functioning of modern scholarship and modern society. As I write this in June 2021, COVID-19 has gone from being an unknown and scary disease to one that has several effective vaccines and is becoming controllable. Developing those vaccines, so quickly, required the full engagement of the global community. Thousands of researchers across the world were able to turn on a dime and develop vaccines based on an entirely new approach—introducing mRNA to tell our own cells

---

[4] And yes, at the time that group was mostly men, but over the years it has become more diverse.

to produce the coat protein as an antigen to fire up our immune systems to produce antibodies. Such miracles would not be possible without an integrated global research community.

13.1 Informal Networks and Colleagues

An important factor in building the integrated global community is our networks of traineeship and "parentage." I'm proud of my academic family tree and to have Mary Firestone as my academic parent and Jim Tiedje as grandparent. I am also friends with many academic siblings and cousins, with whom I generally share scientific vision and approach. We identify as the Firestone clan; now my trainees identify a Schimel clan. Such relationships are powerful and long-lasting, which is why the National Science Foundation (NSF) defines a *permanent* conflict of interest between graduate advisor and advisees—we can never review each other's proposals.

But as we develop connections through meetings and visits, we build extended professional families. These broader networks are also very much part of our professional lives. Aside from reviewing one another's work and drinking beer together at conferences, one of the main ways we interact is through our networks of trainees—we send undergraduates to our colleagues as graduate students, and we pass our graduate students to colleagues as postdocs.

When you have someone extraordinary, sending them on to work with a colleague is easy and a pleasure. Few of our trainees, however, are perfect—rather, they are human and have strengths, limitations, and personality glitches that a potential advisor might want to know about. How much should one say about those glitches? If one of my undergrads asks me for a letter to apply to medical school, I'll be honest about their strengths, but I feel little obligation to dwell on their quirks. But, if that same student were applying to graduate school in a department where they will work with a colleague of mine? That colleague is *also* "my people," and I have obligations to them! How do I balance my obligations to my student and to my colleague?

My motto for mentoring is to help my people get to the place that is right for them, but that can involve a complex equation. I need to be honest with my students about my colleagues' strengths and limitations as potential advisors, but I must be equally honest with my colleagues about my students. I know people who have "blacklisted" colleagues for writing a letter of recommendation that badly misrepresented a trainee or left out critical information; they felt their colleague had essentially lied. That hurt the letter-writer's credibility forever—and not just about trainees. If someone's letters of recommendation are unreliable, should you trust their papers or proposals either?

Balancing obligations to both the people we are writing to, and about, can be delicate. How do you represent a student who is really bright and talented, but has personality quirks that make working with them challenging? Anything negative in a letter glows like red neon and can overwhelm any number of glowingly positive things. So, what to do? You could just vaguely allude to the concern. Then if a colleague were to call you on it later, you could say that you did mention it. But, I'd be more pissed at someone for "lying with the truth" that way than if they hadn't said anything at all—they knew about the issue but didn't tell me?

Well, we still have telephones! And Zoom. When there is a delicate conversation, writing is *not* the way to have it—talking is. There, you can say that "Josh is great, and I think you should take him on, but there are some things that I should give you a heads up about."

This kind of backroom networking can be problematic in making it difficult for outsiders without the "right" pedigree to break into academe—having connections helps. But these networks don't just help the "rich." I invest heavily and personally in my people so I'm risk-averse. I value input such as: "Anita's undergraduate transcript isn't perfect. It took her a few years to figure out college and to find her path. Her overall grade point average isn't great but as she figured things out, her grades got better. She'll need to take some extra classes to fill in some gaps, but that won't be a problem. She was amazing in the lab: dedicated, thoughtful, and creative. By the end, she was fully a peer with the Ph.D. student she was working with—she has what it takes and you'll be happy if you take her on." You can say that in a letter, but it's more convincing face-to-face. And if the issues are personal or delicate, it's more important to say in person.

Supporting and growing our professional communities and networks calls for balancing obligations in several directions: to open and expand our communities, to support our people, and to treat our peers and colleagues honestly and fairly. We need to keep all of these in mind as we interact and work to build academe and develop the new generations of our professional communities.

13.2 Formal Networks: Professional Societies

Beyond the informal networks of academic families and collegiality, the main mechanism that we use to organize our academic lives outside our universities is via professional societies. These are how scholars self-organize to create intellectual homes. Societies organize conferences, confer awards and recognition, establish codes of professional conduct, and publish journals. Society membership and participation carries privileges and benefits.

For all their value, however, academic societies are "commons"—owned by none and owned by all. Organizations don't run themselves. Conferences don't happen by magic and journals don't appear out of nowhere. Every one of these entities require work and commitment from its members to keep functioning: organizations need officers to run them, journals need editors and reviewers, granting foundations need program managers. Societies also need our membership dollars to support the enterprise, and member numbers to give them credibility in representing our communities. You can, for a while, skive off and let others do the work while you harvest the benefit, but if everyone were to act that way, the systems would collapse to the detriment of all. As commons, our professional organizations require communal support and engagement.

I suspect that few of us, when we are starting out, fully understand the real role these entities play in our working lives. I think it sneaks up on us, as I realized when I found myself attending section business meetings at conferences as a matter of routine. My friends and peers were going, so I would, too. Over time, you end up becoming assimilated into community. But it also took time to come to understand how much goes into maintaining the systems that we depend on. An early lesson was when I was a Ph.D. student; Mary came to the lab and told us she'd been asked to serve on the National Science Foundation's ecosystems science review panel—a three-year commitment for two panel meetings a year in Washington, DC, plus reading all the proposals beforehand. She asked us what we thought. We all said, "No, don't do it—we already don't see enough of you." She looked at us all thoughtfully for a few seconds, and said, "You haven't come up with any reasons I hadn't already thought of, so I'm going to say 'yes.'" After she left the lab, we looked at each other perplexedly and said "Why did she ask us, if she was just going to ignore us?"

The reason became obvious as I advanced in my career and, in my turn, was asked to do a term on the same panel. By that time though, it had become obvious to me—you don't say no to the NSF! Not only is it an obligation to serve and support the community, but there is no better way to learn how to write successful proposals than to serve on a review panel. Serving on a panel is a key step in developing a research career in academic science. It doesn't matter that it's a hellacious flight from Fairbanks, Alaska to Washington, DC—you don't say no to the NSF.[5] But as students, we hadn't figured that out. We were

---

[5] There are acceptable reasons for saying "no" to NSF, but just being "normally" busy is not a good one.

like kids being told their mother was going on a trip, and we reacted from our own narrow, ignorant, and selfish perspective—mommy, don't go! I don't blame us for not understanding—we didn't know what a professor's job actually entails. We all learned. Everyone I can remember being in the lab that day has served actively and well. Mary set an example, even if she didn't take the time to fully explain what lay behind her logic.[6] This is what citizenship looks like in academia.

I will note that my claim here that there are some things that you just don't say "no" to might seem at odds with my advice about saying "no" from Chapter 2 on being an Assistant Professor. It's not. You shouldn't say "no" to career benchmarks, important networking opportunities, or valuable learning experiences. Some of these activities involve a lot of work, but they measurably advance your career. You say "no" to the grunt activities exactly so that you have time to say "yes" to these truly important ones!

These community service activities are sign posts that your career is developing, and they affirm your growing stature. They are valuable on your record. They are also usually a lot of fun, working with your colleagues. Importantly, these professional activities are how we support the communities that support us—they are just the larger, system-level version of *take care of the people who take care of you.*

---

[6] Mary wasn't taking a vote; rather, she had paid attention to us, and she had exactly described her reasoning. That she even bothered to ask illustrates why I dedicated this book to her.

14
Resolution

Thriving in Academe

> **Profession**: An occupation in which a professed knowledge of some subject, field, or science is applied; a vocation or career, especially one that involves prolonged training and a formal qualification.
>
> **Oxford English Dictionary**

I love academia. I love knowledge and the processes by which we create it. I love exploring nature and how various systems interact to create organisms, ecosystems, and societies. I am constantly amazed by the magic of our pull-yourself-up-by-the-bootstraps processes of research and scholarship. We are the engine of modern societies, creating critical knowledge and technology, training people in a dizzying array of skills, and preparing people who will teach the next generation. Universities are magical. And the core of that magic lies in the nexus of scholarship and teaching that simultaneously create both today's new knowledge *and* the next generation. So, yes, I love, and even revere, academia.

That doesn't mean that I think academia is perfect—of course it's not. Universities, as I've said, can be quite infuriating places to work. We reflect society and we suffer the ills that plague society—ego-fueled ambition, departmental self-interest and politics, racism and sexism. We also suffer from the underfunding of higher education that seems based on the increasing sense that a college degree confers an *individual* benefit to the student, rather than a communal benefit to all.[1] We struggle to live up to our ideals and our visions, but we at least try, which is more than I'll say for some institutions. Despite our imperfections, academia remains an amazing enterprise and reflects the best of what our society has to offer. At our best, we produce magic.

---

[1] I suspect some of that results from our diversifying student populations. It's easy to support publicly funded college when it's *your* kids going. But when it's someone else's kids? Let them pay their own way. But there's also a feedback loop: as public funding lags, tuition goes up to cover the gap, so people become less supportive of public funding, so funding lags further, so tuition goes up . . . .

But the magic that academia produces isn't produced by magic. Its produced by people! It's produced through the elaborate human systems and structures that allow us to function. As I said in Chapter 7, without the administration and operating staff, I'd be sitting in an empty field pontificating like Plato. Fundamentally, therefore, academe is about people. We may work on nonhuman organisms or nonliving systems, but we work *with* and *for* people.

I ended *Writing Science* with the message that as a scientist, you are a *professional writer*. Here I can drop a word: as an academic, you are a *professional*. I've noted the classic military wisdom that amateurs study tactics, while professionals study logistics: "the detailed coordination of a complex operation involving many people, facilities, or supplies." If that does not describe a university, I don't know what would. Thus, becoming effective as a professional academic calls for developing some appreciation of the *complex operation* and particularly of the *many people*.

To thrive in academe, you have to be excellent at your scholarship and teaching. But just being good at that job isn't sufficient. Thriving and making the most of your career calls for managing the human systems that regulate our lives: our trainees, our academic peers and colleagues (on and off campus), and the people who make our administrative systems function.

There are a few principles to achieving that, but they all come down to some version of *take care of the people who take care of you*.

Successfully working with people calls for effective mutualistic relationships. Some people may be predatory, parasitic, domineering, or autocratic, but such relationships are hard to maintain and often end badly. The "peasants" get resentful. Life is easier when people have your back and support you because they want to, rather than because they're afraid of you. To quote President Truman: "It is amazing what you can accomplish if you do not care who gets the credit."

A mutualistic relationship is one where each partner provides the other something they value. That means understanding what they need so you can provide it—so that they will listen and attend to your needs in return. Our trainees need us to provide them with the intellectual support to develop as scholars, the emotional support to survive the stresses, and the resources to survive; in return, they support us through their scholarship and teaching. Our department mates need us to support them by keeping our departments functioning and advancing. Our staff need us to work within policies when we can and to be sensitive to their workloads and constraints. None of these things seem that difficult on the surface; yet, a relatively limited fraction of academics manage to put the pieces together effectively. Why? Many are content to just focus on their own scholarship. Yet, becoming effective as an academic professional requires some knowledge of how our systems work. Who does what? Who controls budgets and decisions? Does the chair have the ability to do what you want them to?

Universities vary in their balance of top-down versus bottom-up decision-making. For example, in the University of California, our administrative structures may look hierarchical, with chairs, deans, provosts, chancellors, and a president. But actual power—the ability to make decisions, give orders, and have things happen—is remarkably dispersed and diffuse. Academic senate committees can be more influential than the administrator to whose territory they relate. For example, the Dean of Graduate Division oversees graduate affairs—but the policies the dean enforces are created by the academic senate's Graduate Council. Our deans don't allocate new faculty lines to departments—they make recommendations to the Committee on Planning and Budget (CPB), which in turn makes recommendations to the Executive Vice Chancellor. If you want to make things happen, you need to know where power actually sits—would it be more effective to talk to your chair, to the dean, or to volunteer to serve on CPB? Pushing on the wrong lever will take time and energy, but achieve nothing. Which is the right system, and how can you best engage the people within that system?

When we move into a new environment, we need to learn its rules. I have a friend who once stepped outside just to grab the newspaper, but without ensuring that the door was unlocked. In Santa Barbara, that might have been embarrassing. In Fairbanks, Alaska, in the depths of winter, it damned near killed her—you need to know the "rules."

To effectively engage your system, it helps, as well, to have a measure of humility. You may be a great scholar, but others around you are, too, and for our staff support teams, that isn't what defines your value. Even if you do bring in more grant money, have higher citation rates, or get better teaching evaluations than your colleagues, so what? In ancient Rome, at a triumph for a conquering hero, a slave stood behind the great man whispering "memento mori"—remember you are mortal! Some of us would benefit from that reminder.

Being good at your scholarship and teaching is likely enough to earn you the champagne pop celebrating tenure. To take your career to the next level, to become a leader, and to be effective in creating the "next generation academe" that better lives up to its promise, however, simply being a good scholar is *not* enough. To reach that next level takes more. It calls for becoming a true professional—someone who understands your human systems and can work with and support all the diverse people who make them work. Those are the people who take care of you; take care of them so that they will.

APPENDIX 1
Useful Resources

1. Career Advice

Baker, V.L. (Ed). 2019. *Success after Tenure: Supporting Mid-Career Faculty*. Stylus Publishing. Sterling, VA.
Boice, R. 2000. *Advice for New Faculty Members*. Allyn & Bacon. Needham Heights, MA.
Goldsmith, J.A., J. Komlos, and P.S. Gold. 2001. *The Chicago Guide to your Academic Career*. The University of Chicago Press. Chicago.
Haviland, D., A.M. Ortiz, and L. Henriques. 2017. *Shaping Your Career: A Guide for Early Career Faculty*. Stylus Publishing. Sterling, VA.
Reis, R.M. 1997. *Tomorrow's Professor*. IEEE Press. New York.
Rockquemore, K.A. and T. Laszloffy. 2008. *The Black Academic's Guide to Winning Tenure—Without Losing Your Soul*. Lynne Rienner Publishers. Boulder, CO.

2. Writing and Publishing

Germano, W. 2013. *From Dissertation to Book* (2nd ed.). The University of Chicago Press. Chicago.
Germano, W. 2016. *Getting in Published* (3rd ed.). The University of Chicago Press. Chicago.
Hayot, E. 2014. *The Elements of Academic Style*. Columbia University Press. New York. This focuses on the humanities.
Heard, S.B. 2016. *The Scientists Guide to Writing*. Princeton University Press. Princeton, NJ. I'll vote for this book as the second-best on writing in the sciences, and the best that covers all aspects of producing an article.
Schimel, J. 2012. *Writing Science*. Oxford University Press. New York. OK, so I'm prejudiced, but I'll vote for my book as #1.
Sword, H. 2007. *The Writer's Diet*. Chicago University Press. Chicago. This is an awesome little book, and the Word ad-in App is equally awesome—just armor plate your ego before you try it.
Sword, H. 2012. *Stylish Academic Writing*. Harvard University Press. Cambridge, MA.
Sword, H. 2017. *Air & Light & Time & Space*. Harvard University Press. Cambridge, MA.

3. Teaching and Mentoring

Byars-Winston, A. and M. Lund Dahlberg (Eds.). 2019. *The Science of Effective Mentorship in STEMM*. The National Academies Press. Washington, DC.
Fink, L.D. 2013. *Creating Significant Learning Experiences*. Jossey-Bass. San Francisco.
Johnson, W.B. and C.R. Ridley. 2018. *The Elements of Mentoring* (3rd ed.). St. Martins Press. New York.
McKeachie, W.J. 1994. *Teaching Tips*. D.C. Heath and Company. Lexington, MA.
Nyquist, J.D. and D.H. Wulff. 1996. *Working Effectively with Graduate Assistants*. Sage Publications. Thousand Oaks, CA.

Wright, G. (Ed.). 2015. *The Mentoring Continuum: From Graduate School through Tenure*. The Graduate School Press of Syracuse University. Berkeley, CA.

4. Miscellaneous

Christy, S. 2010. *Working Effectively with Faculty*. University Resources Press. Syracuse, NY.

Daniell, E. 2006. *Every Other Thursday*. Yale University Press. New Haven, CT.

Haggerty, K.D. and A. Doyle. 2015. *57 Ways to Screw Up in Grad School*. University of Chicago Press. Chicago.

Ruben, A. 2010. *Surviving Your Stupid Decision to Go to Grad School*. Broadway Books. New York.

APPENDIX 2

Mottoes for Memorable Mentoring

As a last tidbit, I'll end with my "Mottoes for *Memorable* Mentoring." Years ago, I was invited to participate in a panel discussion about mentoring in graduate training. I have several real mottoes that you've seen extensively throughout the book:

- As a professor, it is your job to produce both scholarship and *scholars*.
- As a mentor it's your job to help your people get to the place that is right *for them*.
- Never let the rules get in the way of doing the right thing.
- Take care of the people who take care of you.

But, for that workshop, I didn't want to just be sappy, talking about being a good person, taking care of your people, and so on. So, I decided to have some fun. I came up with a set of mottoes that, if you were to follow them all, would guarantee that you would be *memorable*. Memorable, perhaps, as someone that students would probably nickname some version of Voldemort. *Joshemort*? Not a name I aspire to. Yet, some of these mottoes have a grain of truth at their core, while several others may be completely false throughout, but still reflect attitudes I've heard colleagues express.

Mottoes for *Memorable* Mentoring

1. Remember that we have graduate students for the same reasons farmers have kids: we need the cheap labor.
2. When they finish, your students will become your competitors. Why would you want to do anything that would advance their careers after they graduate?
3. Students need to learn effective time management: *Days have twenty-four hours, don't they? There are seven days in a week: Weekends are so you can get work done without distraction.*
4. Graduate school is so much fun, students don't need vacations. *Corollary: A student's standing should be based on how many hours they are in the lab, not how much they accomplish in the hours they do spend there.*
5. Academia is rough—you will experience harsh and undeserved criticism. Make sure students are used to it.
6. Credit is finite and a zero-sum game. If you give credit to your students, you lose it.
7. Make sure that your students know that when they get their Ph.D., they too will know everything. Until then, they know nothing and aren't worth listening to.
8. Make sure that students know about the "pot of gold" at the end of the road: a job. Don't tell them that you will write a letter of recommendation that will ensure that they don't get it.

Corollary: Let your people know they are wonderful, up until the day you fire them.
I think the first five mottoes are grounded in some measure of truth. I discussed my real thoughts on these issues in Chapters 9 and 10. In the natural sciences, we rely on our lab groups to do the work that carry our names, but that is generally less the case in the humanities. The social sciences span the range of approaches from individual, book-culture, to group-based,

article-culture. When students finish their degrees and become professors, they start applying for grants, recruiting students, and so on. They may well compete for things or people you want—more power to them! Aren't you proud of them? You should be! Although I might have kvetched at the time when I was trying to recruit someone who, instead, went to work with a former student or postdoc, it still gave me a warm glow to see my people doing well (damn them). I also had turned down a grad school offer from Jim Tiedje to work with his recent student Mary Firestone. Its payback time. Graduate school (and scholarship generally) takes a lot of time and personal commitment—but it shouldn't be all-consuming. The final truth is that academe will bring criticism; we've all gotten snarky reviews on papers and had proposals rejected. I'm still annoyed over some things people have said. So, yes, it is important to help students develop the coping skills to deal with such things.

The other points (6, 7, and 8) I disagree with *vehemently*—those I see as having not even a grain of truth whatsoever, even though I have known people who seemed to believe them. My experience is that the more you give credit for the work to the students who did it, the more it also sticks to you—as advisor you get credit for both the work *and* the student—a double win, rather than a loss. If you're going to be jealous of your students, quit. I've had several students who I thought were smarter and more talented than I am; and they've written papers that I am proud to have been part of! It's our job as mentors to be open and honest with our trainees, and to support them in reaching the place that is right for *them*. If they're not performing at the appropriate level, level with them while they can fix the problems, or at least avoid wasting time.

Sources Cited

Baker, V.L., Lunsford, L.G., Neisler, G., Pifer, M.J., and Terosky, A.L. (Eds.). 2019. *Success after Tenure. Supporting Mid-Career Faculty*. Stylus Publishing. Sterling, VA.

Berhe, A.A., Hastings, M., Schneider, B., and Marín-Spiotta, E. 2020. "Changing Academic Cultures to Respond to Hostile Climates." In Azad, S. (Ed.), *Addressing Gender Bias in Science & Technology* (pp. 109–125). American Chemical Society.

Bernhardt, E. 2016. "President's Environment: Being Kind." https://freshwater-science.org/news/presidents-environment-being-kind

Blanchard, K.D. 2012. "I've Got Tenure. How Depressing." *Chronicle of Higher Education*. https://www.chronicle.com/article/ive-got-tenure-how-depressing

Cantor, D., Fisher, B., Chibnall, S., Harps, S., Townsend, R., Thomas, G., Lee, H., Kranz, V., Herbison, R., Madden, K. 2019. "Report on the AAU Campus Climate Survey on Sexual Assault and Misconduct." Prepared by Westat for The Association of American Universities. https://www.aau.edu/key-issues/campus-climate-and-safety/aau-campus-climate-survey-2019

Carver, P. 2017. Women of Color Ph.D. Candidates Thrive in Sister Circles. Diverse issues in Higher Education. https://diverseeducation.com/article/102621/

Chambers, J.Q., N. Higuchi, and J.P. Schimel. 1998. "Ancient Trees in Amazonia." *Nature* 391: 135–136.

Chaudhary, V.B. and A.A. Berhe. 2020. "Ten Simple Rules for Building an Anti-Racist Lab." *PLOS Computational Biology*. 16: e1008210. https://doi.org/10.1371/journal.pcbi.1008210

Daniell, E. 2006. *Every Other Thursday*. Yale University Press. New Haven, CT.

Dryer, D. 2016. "Writing Is Not Natural." In L. Adler-Kassner and E. Wardle (Eds.), *Naming What we Know: Threshold Concepts of Writing Studies* (pp. 27–28). Utah State University Press.

Enfield, N.J. 2017. *How We Talk: The Inner Working of Conversation*. Basic Books. New York.

Firestein, S. 2012. *Ignorance: How It Drives Science*. Oxford University Press. New York.

Finson, K.D. (2002). "Drawing a Scientist: What We Do and Do Not Know after Fifty Years of Drawings." *School Science and Mathematics*, 102 (7): 335–345.

Flaherty, C. 2018. "A Non-Tenure-Track Profession?" *Inside Higher Ed*. https://www.insidehighered.com/news/2018/10/12/about-three-quarters-all-faculty-positions-are-tenure-track-according-new-aaup

Garcia-Williams, A.G., L. Moffitt, and N.J. Kaslow. 2014. "Mental Health and Suicidal Behavior among Graduate Students." *Academic Psychiatry*, 38: 554–560.

Germano, W. 2013. *From Dissertation to Book* (2nd ed.). University of Chicago Press. Chicago.

Germano, W. 2013. *Getting It Published*. (3rd ed.). University of Chicago Press. Chicago.

Glessmer, M., A. Adams, M. Hastings, and R. Barnes. 2015. "Taking Ownership of Your Mentoring: Lessons Learned from Participating in the Earth Science Women's Network." In G. Wright (Ed.), *The Mentoring Continuum from Graduate School through Tenure* (pp. 113–132). The Graduate School Press of Syracuse University. Syracuse, NY.

Harsey, S., & J.J. Freyd. 2020. "Deny, Attack, and Reverse Victim and Offender (DARVO): What Is the Influence on Perceived Perpetrator and Victim Credibility?" *Journal of Aggression, Maltreatment & Trauma*, 29: 897–916.

Hayot, E. 2014. *The Elements of Academic Style*. Columbia University Press. New York.

Hussar, B., J. Zhang, S. Hein, K. Wang, A. Roberts, J. Cui, M. Smith, F. Bullock Mann, A. Barmer, and R. Dilig. 2020. *The Condition of Education 2020* (NCES 2020-144). U.S. Department of Education. Washington, DC: National Center for Education Statistics. https://nces.ed.gov/programs/coe/pdf/coe_csc.pdf.

Institute of Medicine. 2007. *Beyond Bias and Barriers: Fulfilling the Potential of Women in Academic Science and Engineering.* Washington, DC: The National Academies Press. https://doi.org/10.17226/11741.

Jackson, W.M. 2021. "A Black Scientist's Retrospective of His Life in Physical Chemistry." Journal of Physical Chemistry. 125: 5711–5717. https://doi.org/10.1021/acs.jpca.1c05194

Jaschik, J. 2012. "Unhappy Associate Professors." *Insider Higher Education.* https://www.insidehighered.com/news/2012/06/04/associate-professors-less-satisfied-those-other-ranks-survey-finds

Kellerman, J. 2003. *Devil's Waltz: An Alex Delaware Novel.* Ballantine Books. New York.

Kolata, G. 2016. "So Many Research Scientists, So Few Openings as Professors." *The New York Times.* https://www.nytimes.com/2016/07/14/upshot/so-many-research-scientists-so-few-openings-as-professors.html.

Kulp, A.M., L.E. Wolf-Wendel, and Daryl G. Smith. 2019. "The Possibility of Promotion: How Race and Gender Predict Promotion Clarity for Associate Professors." *Teachers College Record,* 121: 050307.

Lamott, A. 1994. *Bird by Bird.* Anchor Books. New York.

Leitner, S., Homyak, P.M., Blankinship, J.C., Eberwein, J., Jenerette, G.D., Zechmeister-Boltenstern, S., Schimel, J.P. 2017. "Linking NO and N_2O Emission Pulses with the Mobilization of Mineral and Organic N upon Rewetting Dry Soils." *Soil Biology & Biochemistry,* 115: 461–466.

Marin-Spiotta, E., R.T. Barnes, A.A. Berhe, M.G. Hastings, A. Mattheis, B. Schneider, and B.M. Williams. 2020. "Hostile Climates Are Barriers to Diversifying the Geosciences." *Advances in Geosciences,* 53: 117–127.

Maron, N.L., C. Mulhern, D. Rossman, and K. Schmelzinger. 2016. "The Costs of Publishing Monographs: Toward a Transparent Methodology." Ithaka S+R. https://doi.org/10.18665/sr.276785

Matthews, K. 2014. *Perspectives on Midcareer Faculty and Advice for Supporting Them.* The Collaborative on Academic Careers in Higher Education (COACHE). Cambridge, MA.

Meyer, J.H.F., and R. Land (Eds). 2006. *Overcoming Barriers to Student Understanding: Threshold Concepts and Troublesome Knowledge.* Routledge. London.

Morgan, T.J.H., N.T. Uomini, L.E. Rendell, L. Chouinard-Thuly, S.E. Street, H.M. Lewis, C.P. Cross, C. Evans, R. Kearney, I. de la Torre, A. Whiten, and K.N. Laland. 2015. "Experimental Evidence for the Co-Evolution of Hominin Tool-Making Teaching and Language." *Nature Communications,* 6: 1–8.

National Center for Education Statistics. 2020. "Race/ethnicity of college faculty." https://nces.ed.gov/fastfacts/display.asp?id=61

NCSES [National Center for Science and Engineering Statistics]. 2019. "Survey of Earned Doctorates." https://ncses.nsf.gov/pubs/nsf20301/data-tables

Rockquemore, K.A. 2016. "Re-Thinking Mentoring: How to Build Communities of Inclusion, Support and Accountability (a webinar)." National Center for Faculty Development and Diversity. https://www.facultydiversity.org/webinars/rethinkingmentoringwk1

Rockquemore, K.A. and T. Laszloffy. 2008. *The Black Academic's Guide to Winning Tenure—Without Losing Your Soul.* Lynne Rienner Publishers. Boulder, CO.

Rucks-Ahidiana, Z. 2021. "The Systemic Scarcity of Tenured Black Women." Inside Higher Ed. https://www.insidehighered.com/advice/2021/07/16/black-women-face-many-obstacles-their-efforts-win-tenure-opinion

Schimel, J. 2012. *Writing Science*. Oxford University Press. New York.

Schimel, J.P. and J. Bennett. 2004. "Nitrogen Mineralization: Challenges of a Changing Paradigm." *Ecology,* 85: 591–602.

Seidel, S.B., A.L. Reggi, J.N. Schinske, L.W. Burrus, and K.D. Tanner. 2015. "Beyond the Biology: A Systematic Investigation of Noncontent Instructor Talk in an Introductory Biology Course." *CBE—Life Sciences Education,* 14: 1–14.

Index

For the benefit of digital users, indexed terms that span two pages (e.g., 52–53) may, on occasion, appear on only one of those pages.

Tables and figures are indicated by *t* and *f* following the page number

academic adolescence, 109–10, 115–16
academic senate, 17, 31–32, 62–63, 75, 157
administration/administrators, 46–47, 50, 61–62, 64, 72, 76, 168
advising styles, 95, 97, 106
Alaska, 10, 68, 94, 168
altruism, 97
ambition, 99, 167
antiracism, 38
arrogance, 51, 68, 78, 88–89
article processing charges, 136
assessment, 55–56, 135
assistant, teaching, 56, 69, 88
Augustine, Saint, 3

battles, choosing, 26, 42, 72, 75
Berkeley Morris, 109, 122
Bernhardt, Emily, 21, 94
Bias, implicit, 38–39
Biogeochemistry, 35
Blanchard, Kathryn, 29
books, academic, 32, 136, 141–42, 145–46
Bren School of Environmental Science & Management, 16
bridges, building, 82–84
bureaucracy, 61

California State University System, 48
career trajectory, 25, 29, 34
Catch-22, 25–26, 101
Centre National de la Recherche Scientifique (CNRS), 71
chain of command, 70, 72–73
chair, department, 15, 35, 49–50, 56, 62–63, 70, 72, 76, 81
Chambers, Jeff, 144
Chapin, Terry, 10
Chicano/Chicana Studies, Department, 65
children, 31, 37–38, 45–46

Chomsky, Noam, 160
Churchill, Winston, 29, 93, 140
citizenship, academic, 51, 53, 164–65
civil rights movement, 42–43
Collaborative on Academic Careers in Higher Education (COACHE), 31
Commander's Intent, 98, 127
committee
    academic personnel, 62–63
    budget & planning, 75
    Ph.D., 64
    search, 8, 10, 18, 19–20
    senate, 168
    space, 18–19
communication, 108, 119, 123–24
competition, 8, 26–27, 157
compromise, 49
conference, 44, 84, 94, 102, 160–61
confidence, crisis of, 102, 107
conflict
    constructive, 41–43
    of interest, 30, 150, 162
content (in teaching), 124–26
Coup d'etat, 72
courage, 100, 107
CV (curriculum vita), 11, 35, 149
criticism, 39, 95, 151
culture, academic, 15, 32, 39, 41, 55, 61, 67, 107–8
curse of knowledge, 54–55, 142, 147–48, 152

Daniell, Ellen, 43
DARVO, 46–47
dating, 9, 14, 49–50
dean, 9–10, 15, 28, 62–63, 72, 76, 168
democracy, 140
department, academic, 9, 15, 20, 55, 64, 74, 160
director, institute, 15, 18–19, 71

diversity, 9, 25–26, 37, 99–102, 161
Don Quixote, 64, 88
Dr. Suess, 31

editor, 135–36, 140, 145, 147, 150, 153
Editor, Chief, 35
Environmental Studies Program, 49, 70–71
ethics, 33, 54, 83
exams, 109, 115, 124, 125, 127–28
expectations
    social, 100, 103
    teaching, 16, 98, 123–24, 126–30
    for tenure, 29–30
    unreasonable, 86–87, 99

facts, learning, 124, 128, 130
failure, 30, 79, 97, 112–15
families
    academic, 20, 160
    biological, 16–17, 27, 37–38, 100, 114–15
fighter jock, 78
find somebody to ask, 80
Firestone, Mary, xv–xvi, 93–94, 96, 108–9, 162
food, 82
friends don't let friends publish bullshit, 95, 143–44, 147
frustration, 9, 27, 85, 106, 110, 113, 140
funding, 4–5, 52–53, 76–77, 156

Gardner, Helene, 48, 126
Golden Rule, 81–82, 140
Good-Fast-Cheap Pick Two, 131
graduate advisor, 80–81, 111
Graduate Council, 65, 168
grant, research, 7, 21, 69, 74–75
Green Ph.D., 3–5
Groffman, Peter, 150
Group, 43
group dynamics, 108
Gulledge, Jay, 94

harassment, sexual, 33, 37–38, 66, 112
Hayot, Eric, 117–18
health, mental, 110–11
Heard, Stephan, 80
Herbert, Frank, 3
Hippocratic Oath, 111
hiring processes, 7–8, 16
Holland, Elisabeth, 45–46, 150

Holy Grail, 38
Homyak, Peter, 6
humanities, 12–13, 15–16, 23–24, 32, 118
humor in teaching, 122

iClicker, 130
Impostor Syndrome, 104
inertia, 56
Institute of Arctic Biology, 72–73
instructional development, 122, 124
Instructor Talk, 121
intellectual struggles, 106
interview, 8, 11

Jefferson Airplane, 80
justice, 62–63

Kellerman, Jonathan, 105
King, Martin Luther, 42–43
Kiss up–Kick down, 88–89

Lamott, Anne, 36
Law of Laziness, 56
leaders, 34, 61, 65, 149
leadership, 26, 46–47, 50, 72
leaky pipeline, 38
learning objectives, 127
leaving a position, 26–28
Lecter, Hannibal, 29–30
lecturer, 48–49, 52
    continuing, 48–49
    with security of employment, 49
letter of recommendation, 30, 50, 162–63, 171
LGBTQ, 9, 37–38
Liebig's Law of the Minimum, 128
Little, Jo, 82
logistics, 59, 75, 167
Lord of the Flies, 5
lubricating relationships, 71, 82–83, 85–86
lying with the truth, 83, 163

manager, departmental, 80–81
Manhattan, 9–10, 86
maternity leave, 45–46, 66–67
McCarty, Jessica, 44
McEnerney, Larry, 39
meetings, 53, 84, 108
Menten, Maud, 44
mentor's intent, 98

Microsoft Word, 69
military, 34, 98, 151, 157, 167
minoritized group, 37, 44, 101
momentum cycle, 23–24, 32
money, 52–53, 61, 76
Murphy's Law, 24, 98, 106
mutualism, 97, 167

Napoleon, 88–89
National Center for Ecological Analysis and Synthesis, 19–20
National Center for Faculty Development and Diversity, 22, 39–40, 102
National Institutes of Health, 25
National Science Foundation, 18–19, 25, 157, 162
negotiations, 14–15, 56–57
network, support, 22, 43, 100, 145, 161, 162–63
news, bad, 87, 110–11
no, saying, 24, 35, 40–41, 164

OCAR, 126
Old Boy's Club, 161
open-access publication, 137, 146
Oswood, Mark, 123
overconfidence, 107

parasite, 97, 167
partner, romantic, 16–17, 45, 54, 112
passion, in teaching, 120–23
pedagogy, 53, 119, 130–31
pedigree, 163
peer review, 135, 140, 147, 149–50
people of color, 9, 25–26, 37, 99–102
personalities, prickly, 85–86
Pirates of the Caribbean, 127
policy, 64–67, 87
politics, 11, 16–17, 18–19, 20, 21, 62, 73–75
postdoc, 45–46, 66–67, 93–94, 112, 115–16
practice, 67–68
pregnancy, 32, 37–38, 45, 66–67
priesthood, 113
Principal Investigator, 156–57
professor
  assistant, 14, 164–65
  associate, 30, 139
  full, 29, 37
  teaching, 49
  visiting assisstant, 4–5, 7

proposal
  book, 25, 138, 143
  research, 11, 18–19, 23–24, 156–57
proprietary systems, 68–69
provost, 76
publish or perish, 135

Red Queen's Race, 4
rejection, 141, 143–44, 158–59
requests, unreasonable, 15, 86–87
respect, lack, 44–45, 51–54
review panel, 52, 156, 164
reviewers, recommending, 139
revisions, 154–55
Rockquemore, K.A., 37, 39–40
role model, 22t, 102–4
rubrics for evaluation, 38–39
rules, Grandma's, 66

sabbatical, 48
salary, 14–15, 16–17, 50, 67
Schimel, David, 160
seminar, 12–13, 120, 123
Smaug, 145
social sciences, 4–5, 15, 23–24
Soil Biology & Biochemistry, 35, 138, 144
Soil Science Society of America, 161
Speed of Academe, 41, 62
Spielberg, Steven, 93
staff, incompetent, 85
start-up funds, 14–15, 18–19
STEM, 4–5, 52–53
storytelling, 126, 131
subscription, 136
syllabus, 123–24, 126, 129–30
Szell, Thomas, 120

teaching load, 118
tenure, 8, 17, 20, 24–25, 29–30, 41, 48, 73, 117
threshold concepts, 124–25
Tiedje, James, 162
Title IX, 112
triage, 151, 154–55, 156–57

unions, 52–53, 66–67
University of Alaska, 8, 10, 18–19, 29–30, 72–73, 83, 104, 129
University of Californa
  Office of the President, 76
  Riverside, 6, 139

University of Californa (*cont.*)
  Santa Barbara, 8, 15, 29–30, 49, 62, 78–79
  system, 32, 48, 52–53, 62, 67, 69, 76, 80, 168

Vietnam War, 42–43
virtuous procrastination, 25, 117–18

Waterloo, Battle, 110
women, 37, 163

work/life balance, 30, 102, 108–9
workload, 105–6
writing, 105–6, 118, 130, 136, 148
Writing Science, 3, 31, 106, 117, 123–24, 125*f*, 126, 149, 167

Yoda, 22
Young Turks, 161

Zimmerman, Eric, 75